Advanced Manufacturing in the Automotive Industry

IFS

ADVANCED MANUFACTURING IN THE AUTOMOTIVE INDUSTRY

Edited by
John Mortimer

IFS (Publications) Ltd, UK

Springer-Verlag
Berlin Heidelberg New York Tokyo
London Paris Tokyo

John Mortimer
IFS (Publications) Ltd
35–39 High Street
Kempston
Bedford MK42 7BT
England

British Library Cataloguing in Publication Data

Advanced manufacturing in the automotive industry
1. Automobile industry and trade — Automation
I. Mortimer, John, *1934*
338.4'56292 HD9710

ISBN 0-948507-32-2 IFS (Publications) Ltd
ISBN 3-540-17137-1 Springer-Verlag Berlin Heidelberg New York Tokyo
ISBN 0-387-17137-1 New York Heidelberg Berlin Tokyo

Phototypeset by Parchment (Oxford) Ltd
Printed and bound by Short Run Press Ltd, Exeter

Acknowledgements

IFS (Publications) Ltd wishes to express its acknowledgement and appreciation to the following publishers/organisations for granting permission to use some of the papers reprinted within this book, and for their cooperation and assistance throughout the production process.

IFS (Conferences) Ltd
35–39 High Street
Kempston
Bedford MK42 7BT
England

British Robot Association
28–30 High Street
Kempston
Bedford MK42 7AJ
England

ISATA
42 Lloyd Park Avenue
Croydon CR0 5SB
England

Gaydon Technology Ltd
Banbury Road
Lighthorne
Warwick CV35 0BL
England

Society of Manufacturing Engineers
One SME Drive
PO Box 930
Dearborn, Michigan 48121
USA

Verein Deutscher Werkzuegmaschinenfabrieken ev (VDW)
Cornelinsstrasse 4
D-6000 Frankfurt 1
West Germany

Japan Industrial Robot Association
c/o Kikaishinko Building
3-5-8 Shiba Koen
Minato-ku
Tokyo
Japan

CONTENTS

1. Overviews

2. Assembly and Welding

3. Sealing and Spraying

4. Inspection

5. Flexible Manufacture

6. Warehousing, AGV systems and Simulation

Preface

THE MOTOR vehicle is 100 years old but developments which have taken place in the last five years are perhaps amongst the most dramatic in its whole history. And, looking to the future, the next 10 years are likely to be equally important.

Indeed, the 1980s have seen a major restructuring of the industry. This is as big as the pioneering efforts in mass production in the USA in the early 1900s, or Europe's introduction of sharp product differentiation in the 1950s, or Japan's more recent innovations in production engineering management.

Most of the early years of the motor vehicle were devoted to development of the product. But in recent years the whole emphasis has switched to manufacture. Several factors have given rise to this.

Competition in the industry has intensified, not least because of the growth in Japanese vehicle production. Japanese automakers have built up their world market share from almost nothing in 1950 to around 28% today. Some studies forecast further growth will take place to the extent that Japanese automobile manufacturers could take 40% of the world market by 1990.

In addition, the Third World is now offering vehicles which can compete with those made in well-established industrialised countries. Not surprisingly, the automobile manufacturers in these regions are working hard to match and surpass the efforts set by Japanese competition.

Thirdly, the oil crisis of the 1970s sparked off a massive downsizing programme in the USA as well as an increase in demand for smaller cars throughout the West.

The Japanese motor vehicle industry was quick to respond to the needs of the US market at this time, and it quickly established itself in that area with so-called 'fuel-efficient' cars.

But the Japanese posed industry on both sides of the Atlantic with a problem. The Japanese, after an early and painful learning process, demonstrated they could manufacture good quality cars at low prices. The combination of the two has done as much as anything to change the face of passenger car manufacture in both the USA and Europe.

Taiichi Ohno, the former vice president of Toyota Motor Corporation, and the man attributed with inventing the Just-in-Time concept, points to the fact that Japan was originally a farming country until about 100 years ago and that the spirit of an agricultural people is still firmly entrenched within the Japanese.

He adds that the Japanese are "not good at production control and tend to try to cover that defect with skills in inventory control."

After the Second World War productivity in Japan stood at one-eighth of that in the USA. An immediate benchmark therefore was to increase productivity eight-fold. And the seeds of future success were born out of those early days when it was necessary to make various types of vehicle by diversified small quantity production. Interesting Japanese firms chose not to follow the US route of low-cost mass production of a single model, as developed by Henry Ford.

"We had to develop our own system to produce low-cost automobiles with as high a quality as European and American cars by raising productivity more than eight-fold," said Ohno. "Otherwise the Japanese automobile industry would not be able to survive. It was apparent that we would not be able to compete with European and American nations in the automobile industry if we followed the same method they did. Therefore our only possibility of surviving in this industry was to develop a unique production system of our own."

Out of this and similar work in other Japanese companies came a new attitude to production. Some people would disagree with Ohno; they would claim the Japanese are good at production control. That they do not over-complicate the production process, but instead simplify it so that it is easily understood and maintained.

Japanese firms in the last 30 years have transformed their manufacturing to the point that for some years now companies have been beating a path to their door to walk round their plants in a bid to find the secret of their success.

In a further bid to discover the secrets of Pandora's box, the automobile manufacturers have teamed up with their Japanese counterparts. At the same time the Japanese automobile manufacturers have shifted the centre of their manufacture away from mainland Japan right into the heartlands of the USA and Europe. They have maintained a momentum aimed at worldwide domination that has been totally lacking by their competitors in the USA and Europe who now seem content to hold onto their present markets, rather than attempt to find new ones.

At the same time the Japanese component suppliers, frequently subsidiaries of the automobile manufacturers, have followed their parents in their trek for new markets in new lands like the USA. At the same time component suppliers in Europe and the USA have lost sales to companies in the Far East, Latin America and elsewhere. These companies have been able to offer huge savings on high quality parts. Not surprisingly, US automobile manufacturers – and others – have come to rely more and more on these parts to help combat the cost advantage of Japanese products, often claimed to be at least $2000 a car.

Despite this effort there are those who predict that 20% of US component suppliers could go out of business before 1995 because of offshore competition, including competition from Europe.

In the face of intense competition the scene in the USA and Europe had to change.

In the USA original equipment customers have been forming joint ventures with overseas manufacturers in a bid to cut the costs of small car programmes. It is estimated these firms, since 1978, have invested no less than $85 billion in advanced technology for manufacture and new products.

But perhaps most important of all attitudes towards manufacture have had to change. In both the USA and Europe the familiar transfer lines have gradually given way to more flexible forms of manufacture. Various systems, like the Comau Robogate technique have been evolved and adopted in a bid to meet the requirement for increased flexibility, increased variety, lower inventories, shorter production times and increased quality. At the same time flexible manufacturing systems (FMS) have become respectable to adopt, though in some cases they have not brought the rewards expected of them.

During this time the industrial robot has appeared. As much as any one piece of equipment, with the exception of the computer, the robot has provided the focus attention. Some would go as far as to say that it has been the engine of change; the catalyst which has helped the industry improve its productivity, and provide the means for improved quality and product consistency.

Developed originally in the UK and Europe the robot was recognised by Japanese firms as means to claw their way back into the forefront of productivity. In contrast to some Western firms which sought to put emphasis on making robots for general use, many Japanese firms developed robots for their own use – only later did they attempt to sell them on the open market when they were fully developed.

But now the scene is changing again. Integration is the key word as the automobile manufacturers look to the computer as a device to control the entire production process, from the conception of the vehicle as a clay model, right through the production of the dies and tooling and machining of components, and out to assembly and test.

The philosophy suggests that data should be input only once: that if that data are correct in the first place there is no room for error later. The move in this direction is spurred on by the computer hardware vendors which see the prospect of increased business in the move to computer integrated manufacture.

General Motors, which is developing its business increasingly outside the automobile scene, has already identified MAP (Manufacturing Automation Protocol) as an important tool that could give it a competitive edge over its Far East competitors. MAP is designed too to reduce the power of the computer vendors, hence avoiding the reliance on one supplier. A secondary advantage of MAP is seen in the increased competition it will bring amongst the computer vendors, to the benefit of prices and deliveries.

But another element in the equation, namely materials, is also taking on more importance. Ten years ago, iron and steel were still the traditional materials from which motor vehicles were manufactured. But the perceived requirement for lighter weight vehicles, and vehicles of higher performance and durability, have opened the door to a new vista offered by so-called advanced materials.

These materials offer new routes to near-to-size manufacture in which the need to remove material is largely eliminated. At the same time, other materials, such as the composites and advanced engineering polymers, pave the way to new concepts of vehicle construction.

The volume-produced plastic car is still a vision that has to be realised - and may never be met. But already new concepts of construction have been explored in which a steel shell provides the support for the polymer-based body panels, drive line and road wheels, and seating.

Available already in relatively small numbers such vehicles, were they to become more widespread, would change the whole face of body-in-white manufacture as it is known today. Such a move would have profound implications for the steel makers as well as the manufacturers of line equipment, including welding guns and consumables.

Most exciting of the new generation of vehicles would have been the General Motors GM80 project, a plastics bodied replacement for the Pontiac Fiero. With industry talk of annual volumes of 250,000 a year the GM80 would have paved the way for high volume manufacture of plastics bodied cars. With other projects in the pipeline industry insiders have claimed that eventually GM would have the capacity to build at least one million plastics bodied cars in the early 1990s. However, for various reasons the GM80 has been cancelled.

The arrival of such a concept would also have a profound effect upon methods of assembly. Possibly, more off-line assembly would take place, much of it right away from the site of manufacture. Suppliers would then deliver complete value-added assemblies allowing the vehicle builders to reduce further their manning levels.

The trend is already discernible with items such as air-conditioning units, but it could embrace other assemblies such as seats, doors, instrument fascias, and so on.

New methods of automatic assembly will appear. The electronics industry has already demonstrated that automatic assembly is a proven technique for small complex components; it now remains to use some of these technologies for much larger components, such as engines, transmissions, axles and so on. The process has already begun but the next generation of engines and transmissions will use more advanced assembly methods. These are already being evolved as the diagram here indicates.

These methods could involve greater use of machine vision and other artificial intelligence systems to ensure first-time-right manufacture.

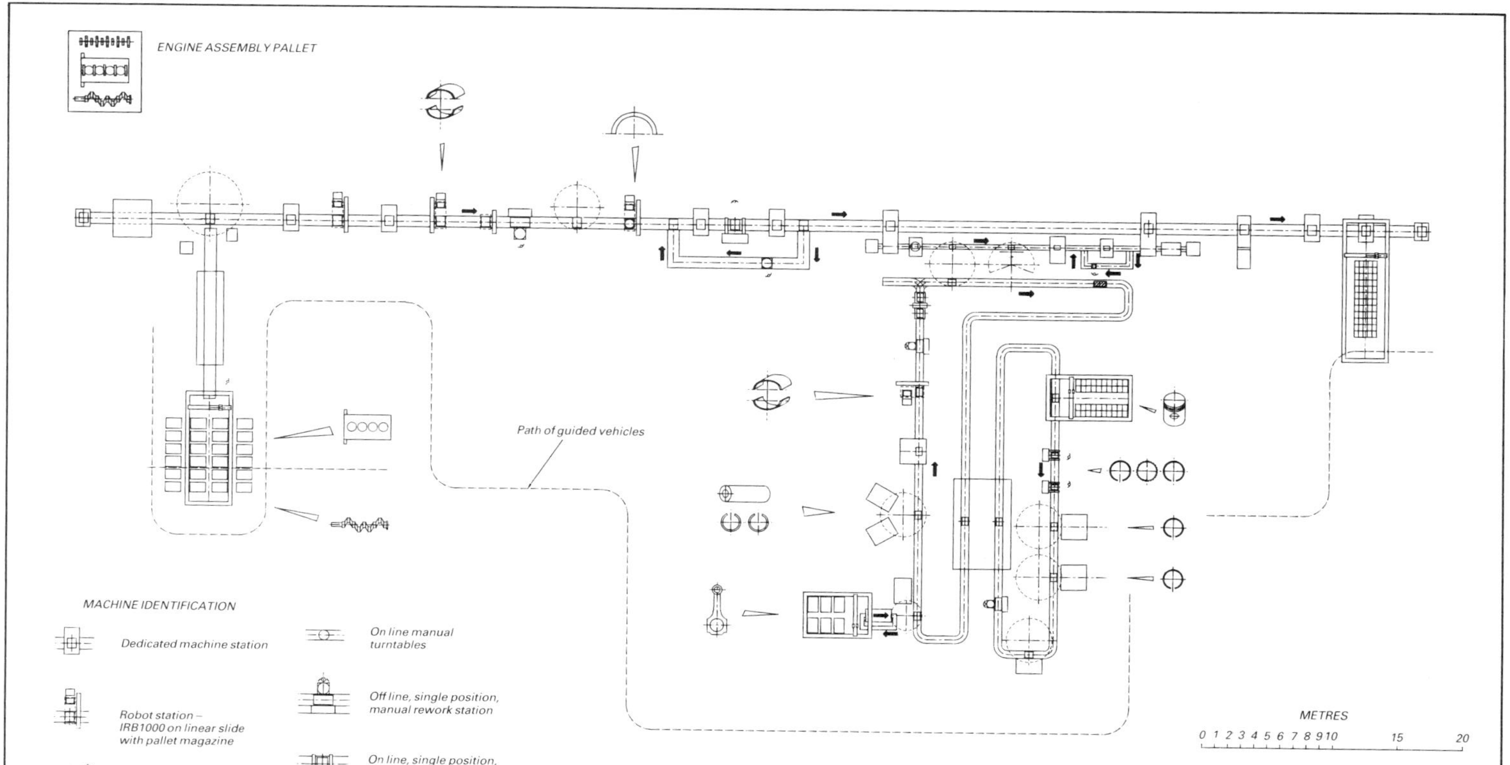

Proposal for first stage of engine assembly by ASEA Manufacturing Systems Ltd.

And the assembly process is likely to become an integral part of computer-controlled testing to provide each vehicle with a quality audit.

Passenger cars have already become more complex as greater use is made of electronic components to add value to the finished product. This trend will continue, opening up new avenues for advanced manufacture of individual components. However, at the same time assembly of the final vehicle will become more complex and time consuming, forcing manufacturers to look increasingly at more off-line assembly.

Already the industrial robot and the automated guided vehicle (AGV) have become common sights in the factories of vehicle builders and their component suppliers. European manufactures of AGVs have been in the forefront of technology and have done much to advanced the use of these vehicles throughout the motor industry. This trend will continue to develop as new forms of guidance are developed. These systems will be integrated with other forms of intelligent materials handling technology and advanced manufacture, including vision and laser guided robots for complex tasks such as seam sealing, arc welding and assembly.

The laser too will become more widespread in its application, alongside such other novel techniques such as water jetting of fabrics and polymers. The laser is a much under-valued technology with much greater scope for application. Heat treatment, cutting, welding and inspection are all areas of application which have not yet been fully explored.

Managers and engineers require to be more inventive and imaginative in the search for cost-effective manufacture, assembly and test routines if their companies are to achieve the status of world-class manufacturers. The divisions between the various engineering disciplines have to become more blurred at the edges as one group increasingly must interface with another.

This is becoming increasingly important in another new technique which is unfolding, namely simultaneous engineering.

World-class manufacturing demands that the relationship between supplier and customer become much closer. In Japan the two work as a team, but in the USA and Europe the relationship is not quite so harmonious. But changes are taking place.

Simultaneous engineering encourages the supplier of machines, tools and components to be involved right at the moment when a new product is conceived. Thus the manufacturing route is developed alongside the product and there are no nasty surprises in store for either party at the end of the road.

The classic area where disharmony can occur between customer and vendor is in engine manufacture. Engines are designed by specialists, all too frequently giving scant attention to the needs of production. Once the engine has been designed the transfer line vendor is involved. The vendor frequently has to incorporate design aspects which would not be required were it to be involved at the design stage.

Simultaneous engineering has important implications for customer and vendor alike. At the least it paves the way for the vendor's production processes to become an integral part of the customer's production. But in the longer term the vendor becomes part of the world-class manufacturing concept that vehicle builders have to survive for if they are to survive.

But the future is more than about mere survival. Survival implies merely maintaining the status quo - clinging onto the edge of a precipice.

Both US and European vehicle builders and component suppliers have to regain the leadership role they once enjoyed if they are to survive in an intensely competitive market place. They must re-establish their own position in the market place - a position that increasingly will centre as much on a clear understanding of the overall objectives as it will on the development and application of technology in component design and in manufacturing to achieve excellence; and the motivation of all employees with a common goal.

Leadership comes from total commitment to the concept of world-class manufacture; to innovation, not for its own sake but for the enhancement of product and manufacturing route; to quality and the belief that the customer is always right; and above all to marketing and the understanding that a determination to succeed is half way to selling the product.

This book shows how many companies throughout the world are tackling the problems of introducing automation in their search to become world-class manufacturers. It is the first time that such an important collection of papers has been put together on the subject.

It is now recognised that achieving the status of world-class manufacturer is prerequisite to having a seat round the table of those vehicle builders which will dominate the scene in the next decade.

Advanced Manufacturing in the Automotive Industry covers the entire spectrum of the manufacturing scene from body-in-white manufacture to final assembly. The papers have been written by acknowledged specialists who have a thorough understanding of their subject. The book is essential reading for all those who wish to learn from others in order to improve their own production processes and move on to develop new techniques and technologies of their own.

John Mortimer
October 1986

1

OVERVIEWS

The management of technological change and the legal implications and considerations when buying factory automation equipment are two important factors with which senior managers need to familiarise themselves. It is all too easy for managers to rush into new technology without first assessing the full implications both in terms of personnel requirements, the profitability of the business and the interaction with other equipment in the company.

THE MANAGEMENT OF TECHNOLOGICAL CHANGE IN THE VEHICLE INDUSTRY

R.A. Dover
Land Rover Ltd, UK

The vehicle industry continues to lead in the application of technology. The reasons for this are examined, and examples given of Land Rover's approach to the increasing pace of change. This paper analyses lessons learned from the successful implementation of change.

It is worth recalling the need for technological change. After all, such change is usually expensive, often disruptive, and occasionally very high risk. No one in industry is making change for the pleasure of spending sleepless nights and long weekends peering towards a disaster on the following morning.

The need for change is driven by the need to satisfy customers and enhance quality. Land Rover is working in an environment of abundant customer choice, and of over-capacity in the industry. It is an environment of constant product development, and necessary cost pressure. The company markets and sells its products in 120 countries worldwide; and apart from a few protected European markets, Land Rover is competing head on with Japanese products.

Although Land Rover's market is broader than most, there are reasons for the explosive growth in the motor industry of robotic applications, CAD/CAM, and continuing high levels of investment in improving production techniques. The costs of distributing and selling products have remained relatively constant. The cost of marketing has increased as a result of over-capacity in the industry and this has eroded profits. It is on the design and manufacturing organisations that the weight of competition pressure largely falls.

The management challenge is therefore planning and implementing change - quickly, effectively, and at low cost.

The risks taken in deciding the level or cost or extent of change depends on each particular environment, and each organisation's business circumstances - on where you are and where you have come from - but the management challenge must represent a high degree of common ground across all of industry.

The key changes

Some 12 months ago Land Rover's manufacturing occupied 14 factory sites, mainly in the West Midlands, including machining, heat treatment, paint, and final assembly operations. Today, these activities are undertaken on a single site at Solihull, in the West Midlands.

In the past 12 months Land Rover has physically transferred or rearranged 60% of the manufacturing facilities. Output per square foot has been improved by 20% and the company has saved approaching one million miles a year of inter-factory transport. The jobs of over 3500 employes have been relocated and energy and rates savings total £2.5 million a year.

Production has not been lost during this rationalisation - and in fact production line rates have increased three times. One new model was introduced and a major facelift was done on another at the same time.

Two examples of the changes made are given below, together with some of the key messages and lessons for success that are being applied to current projects.

The manufacture of pulleys

Part of Land Rover's market is for bespoke products. In addition to the standard production Land Rover, the bespoke offerings include fire engines, ambulances, snowploughs and military vehicles. The company has a major competitive advantage over the Japanese in that it can make relatively easy changes to the bodies, but the running gear - chassis, axles, power unit, and gear train - for all of these vehicles is largely identical. One key difference, though, is that the military vehicle has air conditioning and special electronics.

In order to meet these individual requirements - air conditioning pumps, power steering pumps, alternators - the company produces a large number of turned pulleys at low volume. This would appear to be a classic case where CNC turning equipment would be a logical choice. CNC machining would offer benefits of reduced set up, reduced inventories, reduced tooling and shorter lead times. The factory producing pulleys was closing, so potential savings could be made on layout, real gains made from a reduction in floor space, and the large number of machine tools used for current processing could be avoided.

It was clearly possible to make out a good case for finding a new CNC set up. But management is about making choices - in this case ensuring that the current priorities were applied for the funds available. A different approach was taken. Firstly, the company looked at variety and for product simplification. Every different pulley on file was laid out physically, and with the product design engineers, a simplified product line-up was agreed on. It was possible to standardise on belt sections, and to eliminate a number of components. Every part number created represents a significant cost - not only in its creation, but in drawings, reproductions, spares, stocking, workshop manuals, and in inventory.

By simplifying the problem before making the decisions, it was possible to specify simpler machines and make decisions accordingly. The essential message was to simplify the product before the analytical work was carried through.

Much the same logic was applied to process operations. Could operations be reduced? Could operations be eliminated? Could material specifications be changed, commonised or eliminated? The same was applied to interprocess work. With a reduction in handling or in movement of components, Land Rover has achieved some major successes.

This approach is not always possible - the customer may be demanding an enhanced process or product; examples of this may be in paintwork or in protection from corrosion - but in general it was found that making a review of product and process complexity was a valuable first step.

Assembly and test of engines

Land Rover needed to close the factory where the Rover V8 engine was assembled and tested. There was a conventionally built facility and one option was simply to pick up that facility and relocate it in Solihull.

Land Rover had already gone through a major study on simplification of the product, with good results and future plans to further rationalise the range. A number of derivatives of the V8 engine were built, from 125 bhp to 190 bhp, for a range of customers as well as for internal use. Currently, the company supplies Austin Rover, Morgan and TVR. It was able to reduce its range to seven basic engine builds, with 49 variants covering LH and RH power steering, air conditioning, and emission control layouts.

Most variability was added at the engine dress stage, after the basic power unit had been through test.

A number of alternative solutions were examined: a small multi-functional team was set up to investigate the latest ideas throughout the European industry. The company made its selection from one of the higher cost routes available - and on the basis of future flexibility, selected an AGV system.

It was known, from experience, that engine specifications were likely to be more demanding as the markets split into emission control and non-emission control areas. And, given the projected life of the V8 and its considerable development potential, Land Rover knew that future flexibility would be the key.

How many of us could have predicted 12 months ago that world oil prices would halve? Or for that matter that, while energy prices were plummeting, the price of electricity in the UK would actually rise? The point to be stressed here is that in putting down facilities with a given life, during the period of that life there is a degree of uncertainty, and change should be planned flexibly if at all practicable.

Land Rover had never installed an AGV system before, and many of the potential suppliers were not able to meet the initial specifications drawn up.

The management of this particular change was tackled by creating a team which included not only production engineers, but representatives from the vendor, supervisors, and operators from the current facility. The company kept everyone concerned involved and informed, and on jig development the operators and supervisors made a major contribution.

Before closing the old site, a short track was installed and, using the team that had developed the project, Land Rover not only trained the employees on the hardware, but also explained the software and the philosophy behind what had been done.

Land Rover's AGVs are controlled by a four-wire system with full two-way communication to the supervisors host computer. The high work content variants are routed through sidings ('specials') on the slide.

A three-phased move ensured continuity of production and avoided the need to stockpile engines. All the deadlines were met and the early emphasis on planning and training ensured a smooth transition.

Production reached 100% of line rate within four weeks which is a tribute to the close cooperation that existed between Land Rover engineers and vendors throughout the planning and installation periods.

The flexible system allows production volumes to be varied, or different models to be produced. Another result of the new working methods with longer work cycles has been a reduction in the monotony levels traditionally associated with repetitive assembly work and an increase in overall efficiency. Further job satisfaction has resulted from making each operator responsible for the quality of his or her own work.

For enhanced management control, the information presented on the colour monitor in the supervisor's office could be relayed to the production director or any other management personnel.

Although the new systems have been operating for only a short time, significant benefits are already apparent, including higher quality through improved working methods, 6% reduction in hours/engine, and a lower work-in-progress inventory from over 400 engines previously to less than 200 engines in Solihull.

So far the key messages from this example are: plan for uncertainty, involve the users and get the users to 'own' the problems.

There is one further point which should be brought out, and that concerns timing. Planning the investment and planning flexibility have been referred to. Planning flexibility must include the time dimensions: it may involve spending more, it may involve spending more slowly on a modular type system (e.g. AGVs), it may mean working with temporary methods until the market and customer view is clear. Timing is a key factor.

Concluding remarks

The important points of Land Rover's views of planning technological change are listed below:

- Simplify the product or process.
- Add variability as late in the process as possible.
- Provide only the necessary controls and do not over-organise.
- Concentrate on the end objective and not the method.
- Integrate the design and manufacturing operations.

When selecting options:

- Go for flexibility if at all practicable.
- Vendor participation.
- Get your people to participate, to understand the environment and to own their part of the process.

Remember the end objective:

- Quality products.
- Cost competitive.
- Meeting customers' needs.

LEGAL CONSIDERATIONS IN PROCURING FACTORY AUTOMATION EQUIPMENT

F.M. Greguras
Fenwick, Davis & West, USA

It is argued that some strategic planning for implementing robots and other factory automation equipment in the automotive industry must be complemented by carefully negotiated legal arrangements.

Robotics and other factory automation are a competitive necessity in today's global economy. With only a few exceptions, however, US manufacturing industry has not yet made the technology investments and structural changes necessary to bring about the factory of the future. Two recent studies indicate that some US manufacturers are beginning to make substantial investments to meet this need.

A study released in August, 1985 by the University of Michigan and the Society of Manufacturing Engineers (SME) estimated that annual domestic sales of robots will increase from $240 million in 1985 to $637 million by 1995. Other studies predict a faster sales growth.

According to the Michigan/SME study, the automotive industry will continue to increase its purchase of robots, but its share will decline from 51% in 1985 to 26% in 1995 because of greater sales to other industries. Purchases by electronics companies are estimated to grow from 8 to 16% by 1995.

The US Census Bureau recently conducted a survey to identify the growth of robot use in the US but did not count the number of robots already in use. The Robotics Industries Association (RIA) estimates the number already in use to be between 16,000 and 17,000. The Census Bureau survey found that 5535 devices valued at $357,744,000 were shipped during 1984. The RIA estimated US production at 5136 units during 1984, with a value of $332,500,000.

According to the Census survey, welding, soldering and brazing jobs were the most common industrial functions for robots produced during 1984. Second was the broad classification of educational, hobby and experimental robots followed by materials handling, product assembly, spraying, painting and glueing, and machine tool loading.

Today's implementation of this robotics and other factory automation equipment must be shaped by a vision of the factory of the future. Otherwise, expensive solutions will be short-lived and not part of long-term strategic planning. Integration of various robots and automation systems and approaches will be difficult. Planning must address equipment and software interfaces and standardisation issues because the factory of the future will be a mixed-vendor, multi-system environment.

The role of flexible manufacturing systems (FMS) in long-term planning raises new issues. Software control is a critical element in an FMS and most companies have not used software directly in their manufacturing operations. This shortens set-up times and simplifies customisation of a production line resulting in a faster response to market demands.

Strategic planning for procuring robotics and other factory automation must be complemented by sound legal arrangements. The key legal provisions covering the procurement as well as the implementation and operation of such equipment must be meticulously documented in a contract.

General approach

A procurement should begin with the creation of a team which includes people with business, finance, technical and legal skills. The team must do its homework in advance. Communications and device interference standardisation developments and trends must be a part of business decisions. Vendor and product track records in the marketplace must be diligently investigated. Financial alternatives, including lease arrangements, must be evaluated.

The procurement team's objective should be to acquire a factory automation solution, not merely parts of it. A total systems or 'turn-key' approach should be used rather than just buying pieces of a purported solution. The total system approach is likely to require a multi-vendor procurement, but one vendor should be held responsible for making the system work. As indicated, software will play a significant role in the solution.

The lawyer on the procurement team usually should prepare the agreement. Since the total systems approach will not be reflected in such a document, the vendor's standard printed contract should be used merely as a starting point for drafting and negotiations. The vendor's promises, both verbal and in correspondence, must become part of this written agreement.

Procurement documents

In most procurements the team should issue a solicitation document to selected vendors, a request for proposal (RFP). An RFP is intended to solicit a customised response to the purchaser's specific needs.

The request for proposal, proposal specifications and relevant correspondence must be carefully reviewed and incorporated into the agreement as exhibits. These exhibits are an integral part of the agreement and must be agreed upon and attached to the agreement prior to its signing.

The RFP should describe the vendor's obligations in detail including:

- Detailed specifications for the system to be implemented, both equipment and software, including expandability and compatibility requirements.
- Deliverables, including software, documentation and technical data.
- Delivery schedule.
- Installation and training obligations.
- Acceptance criteria and tests or methodology for developing them.
- Maintenance obligations.
- Managing other vendors' products and people.
- Governing legal terms and conditions.

The statement of the system specifications is critical. They must describe the functions and technical requirements (such as positioning, acceleration and velocity capabilities) of the system in sufficient detail so they can be used to evaluate the vendor's performance. Communications and device compatibility and expandability are key elements of such specifications. Specifications should be developed by the procurement team, in some cases with the assistance of outside technical help. If the vendor participates, the purchaser should retain the right of final approval.

Contract terms and conditions

The RFP should specify the major contractual terms and conditions that will govern the procurement. Presenting the key contractual provisions in the RFP will avoid surprises later which, at a minimum, could delay the procurement.

The relationships between system acceptance, warranties and maintenance must be carefully addressed. An exhibit to the contract should specify the acceptance test procedures to be followed or the methodology for developing them. The purchaser should have the right to review and approve all test results. Both parties have a problem if acceptance criteria are too simplistic or vague.

Performance standard

Once acceptance occurs there should be a system performance warranty period or a maintenance period. The level of performance under either should essentially be that system operation meets the specifications. The vendor's obligation should include the individual units of equipment as well as the entire system.

The traditional difference between a warranty and maintenance is that under a warranty the vendor has the obligation to repair or replace, usually at no additional cost for either service or parts, while maintenance can be the same level of system performance but at an extra cost, at least after some initial period. If the contract price includes a period of maintenance performance equivalent to warranty performance then the labelling is unimportant.

A vendor will start usually by offering warranties only as to the performance of individual components of the system, normally that a unit of equipment is free from defects in materials and workmanship under normal use for a period of x days. The purchaser should require a level of performance related to the operational specifications including down-time measure. The level of performance required can be absolute or substantial conformance to the specifications. The level of system availability without disruptions in operations, a time measure, is a crucial system requirement. The methodology for measuring this must be carefully determined. The vendor's response requirements should be related to the critical nature of the specific problem.

The duration of such warranties is a further issue. The issue is less crucial if the contact price includes a meaningful maintenance obligation. Illustrative periods for a warranty include:

- Only at the time of delivery or installation.
- X days following delivery, installation or acceptance.
- During the term of the agreement (less likely to obtain).

The actions of the purchaser which may void the vendor's obligation should be carefully reviewed. They should be acts which make it extremely difficult or impossible for the vendor to perform. One common exclusion is when the purchaser has modified the system in an unauthorised way. For this reason, the purchaser should understand at the outset what is authorised and be sure that this restriction is not an operational problem.

Vendor's liability limitations

The vendor, like the purchaser, will desire a limitation on its total liability under the procurement. Some limitations on economic losses are commercially acceptable while others are not. In general, limitation of liability provisions are effective only as to economic or commercial losses, not personal injuries.

Some proposed limitations will be more subtle than others. The following are some examples:

- Limitation of express warranties in duration and scope of obligations to those expressly stated in the agreement (commercially acceptable if express warranties are carefully drafted).

- Exclusion of implied warranties of merchantability and fitness for a particular purpose (commercially acceptable if the express warranty is carefully drafted).
- Repair, replacement or refund in the event of a breach as a sole and exclusive remedy (these are red flag words) which means the purchaser may not take any other actions against the vendor (usually acceptable).
- Ceiling on money damages (usually unacceptable).
- Exclusion of certain types of damages such as lost profits and consequentials (usually acceptable if carefully drafted).

Payment and pricing obligations

Payment incentives and/or holdbacks should be used to motivate the vendor to perform fully and on time. The timing of initial payments should be based on successful action; that is, delivery or acceptance, rather than merely receipt of an invoice.

A most favoured pricing provision should be used in a major procurement. Such a provision requires the vendor to give the purchaser the best pricing terms offered to any customer now or in the future. This provision can include 'most favoured' contractual terms as well, such as warranties. This must be drafted very carefully since its intent can be rendered meaningless by words which limit the vendors obligation to arrangements in which the same or similar quantities are purchased under 'the same circumstances'.

If the procurement is ongoing, the purchaser should negotiate for a period during which purchase prices may not increase with a price escalation restriction on increases after the stated period. The vendor's price should include cabling, interconnections between the system components and other 'extras' such as training and installation which are essential to the implementation and operation of the system.

Non-infringement warranties and indemnification

The vendor must warrant that it owns the proprietary rights (patents, copyrights, trade secret) to the technology embodied in the factory automation system (including software) and indemnify the purchaser for losses if a third party claims otherwise. Because of the increasing sophistication of such technology and the importance of software, this contractual provision requires more attention than for previous procurements of traditional fixed equipment. The purchaser should seek the following:

- A broad scope of types of claims covered including the most likely source of actions; that is, patent and copyright infringement, and trade secret misappropriation.
- The vendor's indemnification obligation should begin when a claim is made, not when an award is made by a court.
- The vendor should have the obligation to make at least a reasonable

effort to obtain for the purchaser the right to continue to use the tainted component.

- The vendor's liability should not be limited solely to refunding the purchase price. Again, the words 'sole and exclusive remedy' should raise a red flag.
- Damages for this breach of warranty should not be included within a general limitation of liability provision; that is, to the purchase price of the component in question.
- Any actions that void the vendor's warranty obligation should be carefully negotiated. The warranty should not stop because the system is based on a combination of products provided by several vendors.

Force majeure

The vendor will propose a 'force majeure' or 'act of God' provision which establishes a legal excuse for non-performance under certain circumstances. The contract should carefully define the triggering events and it should protect both the vendor and the purchaser. The purchaser should limit the definitional scope of such circumstances since it has no legal remedy if such an event occurs. In addition to major disasters, a vendor may try to include death, illness, disability of key employees, labour disputes, etc. The purchaser should reserve the right to terminate the agreement if a defined event extends beyond a specified period.

Software contracting issue

As indicated, software will usually be part of a total systems procurement. While the equipment itself will be purchased (or at least in some cases), software is almost always licensed. Unlike full ownership, a license is only a right to use the software in specified ways. The scope of this license must be carefully reviewed since a standard form licence may limit the rights to use the software in unexpected ways. The use rights must be adequate to meet operational needs.

The vendor may propose to label its software as a 'trade secret' and that the purchaser keep the software confidential with a standard of care such as 'strictly confidential'. Initially, the purchaser should carefully define exactly what it is that is confidential and establish the standard of care as its own internal standard for securing its own confidential information to avoid having two standards to administer.

In addition, the purchaser's confidentiality obligation should terminate upon certain specified events; for example, if the subject matter information of the obligation is:

- In the public domain at the time of receipt by the purchaser or comes into the public domain thereafter through no contractual breach by the purchaser.
- Known to the purchaser prior to disclosure.

- Independently developed by the purchaser.
- Disclosed or used by purchaser after x years of receipt.

The purchaser can acknowledge that the vendor claims the software is a trade secret without giving away too much. The purchaser should also consider whether obligations of confidentiality should be mutual since in almost all instances the vendor will be privy to inside information and know-how concerning the purchaser's existing and future products and production objectives.

Concluding remarks

Sound strategic planning for implementing robotics and other factory automation must be complemented by carefully negotiated legal arrangements. The complexity of a total systems, multi-vendor procurement requires a teamwork approach. Successful procurements require careful attention to detail or a purchaser may end up with a group of high-performance components which, when connected, produce unsatisfactory results.

2

ASSEMBLY AND WELDING

The steel-bodied passenger car is likely to continue in production for many years to come, but with increasing emphasis being placed on flexible assembly, greater product variety, and higher standards of quality. The industrial robot is likely to feature heavily in future manufacturing programmes since they have already established themselves as important engines for change. Increasingly, they will be integrated with other advanced automation equipment to broaden the scope of their applications. Important work has already been done to introduce robots into the spot-welding environment, arc-welding and assembly cells. Equipment is now available to allow robots to tackle more complex arc-welding jobs, thanks to the harnessing of the industrial laser which, with machine vision systems, are likely to see wider application in the automotive industry.

LAND ROVER ROBOT WELDING LINE

T.E. Hanley
Land Rover Ltd, UK

In 1981, Land Rover installed a production system consisting of 16 robots to arc weld the chassis for the Land Rover 110 model range. Apart from the anticipated manufacturing and quality improvements from such a system, the objective was to give the company direct practical experience of robot manufacturing techniques. Land Rover operation, in comparison to volume car producers like Austin Rover, is much smaller and the robot welding techniques used at Austin Rover, mainly being spot-welding, could not be directly transplanted into the operation. Several problems were encountered initially, largely due to the progressive move of linking several robots into a fully automatic welding system. This paper describes the work that was carried out to overcome these problems and how new techniques are being introduced to enhance the system further.

The production system was installed by Land Rover Ltd to automate the arc welding of chassis for the Land Rover 110 model range. Fig. 1 shows such a chassis.

The system consists of 12 robots connected by an automatic handling system, forming the main production line, the schematic layout of which is shown in Fig. 2. This line is supported by a further four robots which are used for spares, sub-assembly welding, development work, and training.

The system was implemented with the introduction of a new product and was aimed at reducing the previously high labour intensive methods used for chassis welding and to improve the quality of the final product.

Original design

The original design was based on a transfer line concept to produce two derivatives of the chassis. It was based on known manual welding capabilities, which, at that early stage of automation development, were the only experiences available.

The initial action was to carry out a considerable number of trials

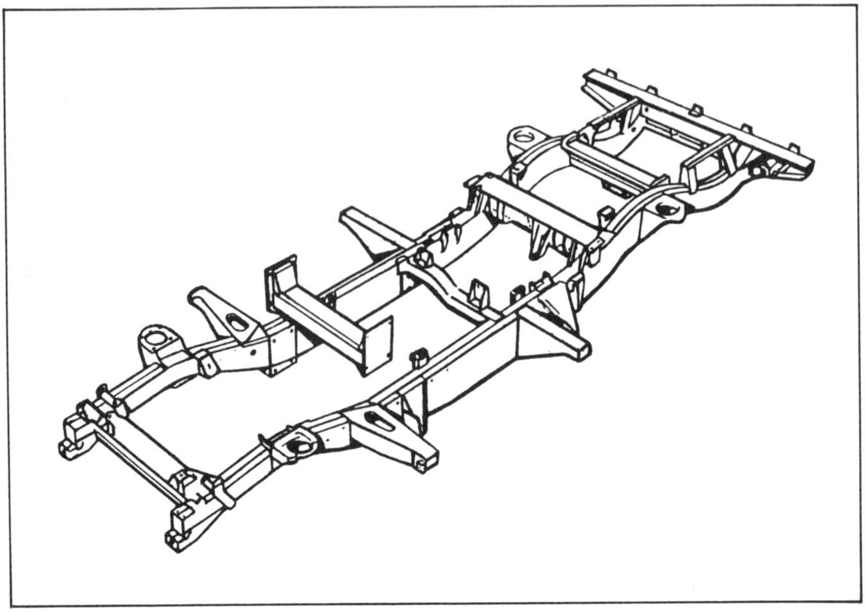

Fig. 1 Chassis for Land Rover 110in. model range

with robots and components to determine the feasibilities and requirements, such as weld angles and access. This also determined which components were to be welded off-line and which, if any, were to be outsourced. These design and installation steps were carried out in conjunction with external consultants and contractors.

Following the initial installation, work has been carried out to overcome a number of teething problems and a continuous programme of refinement has been undertaken to enhance the system as technology has progressed. These developments are outlined under the following headings: tolerances, programming, control system, weld process, robots and jigs, installation/area, back-up spares and continuing development work.

Tolerances

In the early stages, it would be fair to say that nobody had any concept of the extent of tolerance problems that would follow with this type of application.

It was agreed in the design period that all sub-assembly jigs would act as setting fixtures with close tolerances being dictated by the designer. However, in practice this proved impractical. Distortion caused by the welding operation led to jamming of various components within the jigs. A solution was achieved by a combination of improvements to the quality of pressed components and a relaxation of jig locations to the full extent allowed by the main production line.

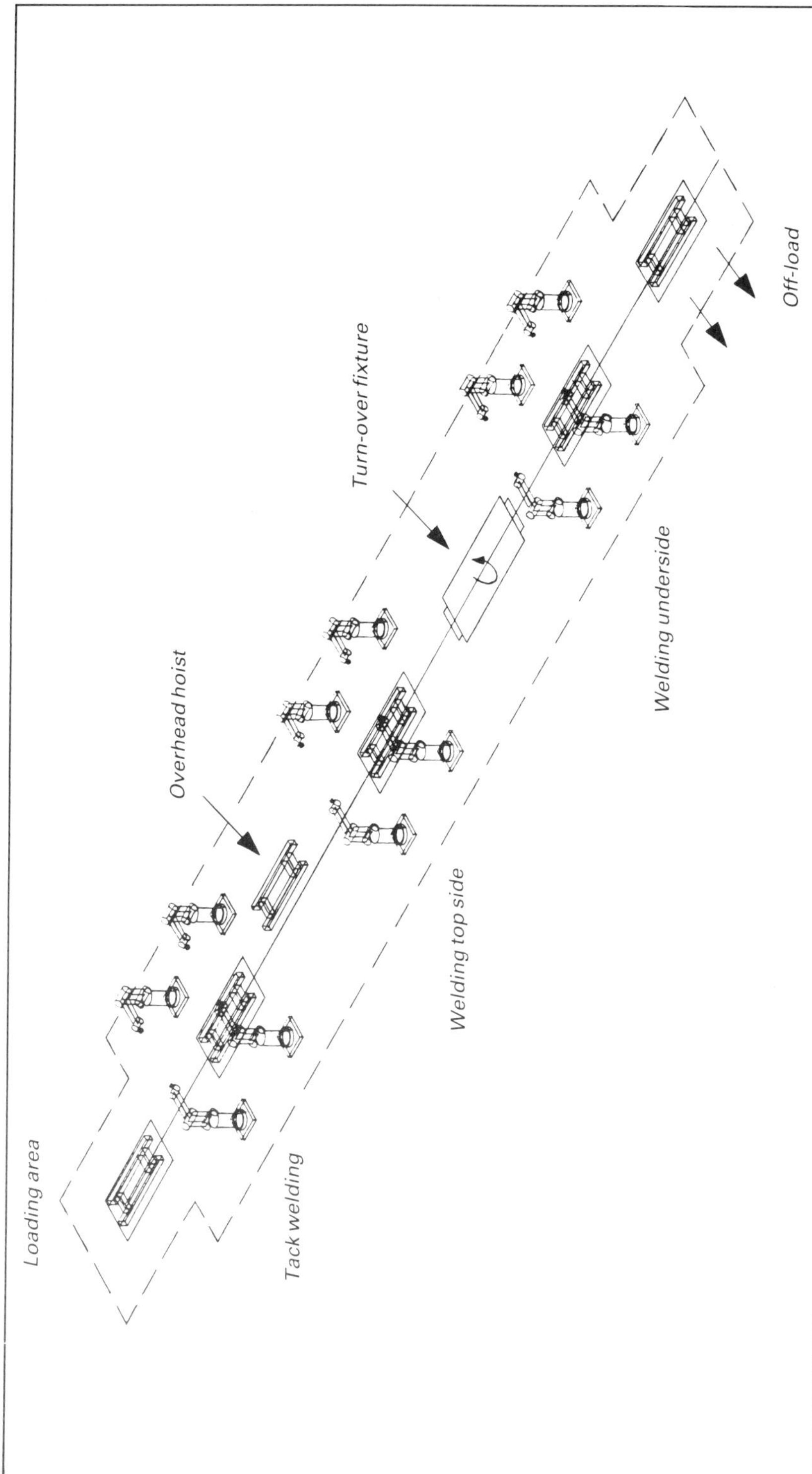

Fig. 2 Chassis auto weld line

The important point here is that tolerances must be determined not only by the required quality of the end product, but by the capabilities of the process itself. The requirement for closer tolerances has placed great demands on the press-shops which are having to improve quality drastically for automated systems. But to achieve these requirements is likely to mean a change of practice where they themselves have to employ new techniques and equipment. This sets a significant challenge to press-shops since customers will not be prepared to accept increases in costs.

The final point with respect to tolerances, is that the labour force traditionally has handled components roughly in this type of environment. This can result in distortion and a loss of tolerances which, while acceptable for manual operations, cannot sustain robot welding. Improved handling techniques were introduced at Land Rover and are essential in any automated welding application.

Programming

Since installation several lessons have been learnt about the programming of this type of system but the main ones are that quick fault-finding and correction are essential. In the Land Rover system, should one robot stop in mid-work then the complete system must be stopped to repair it. If this takes longer than seven minutes to correct then a complete chassis is lost.

The programming of the Land Rover system allows a robot to be stopped without the whole system shutting down, i.e. if it is a case that it simply fails to move from its rest position, then only its own operation will be missed, would the rest carry on. However, if it stops in mid-work, then the whole system would have to stop as the robot arm may well impede the overhead hoist when trying to remove the chassis to the next station. The only time that a robot halts the whole system (by failing to start) is in the first operation where the chassis is tack-welded together, i.e. if it is not tacked together, it will fall apart when the hoist attempts to lift it out.

A major improvement introduced by Land Rover was the introduction of a second run at the weld if the first attempt at starting failed. Even though only two attempts are made before the system is stopped to find the fault; this has resulted in a 40% reduction in weld failures due to the whole process being able to continue without maintenance personnel having to intervene.

It is important also to have a regular check and, if necessary, correction of the zero position of the robot head. If this does not take place, bad welds can result and complete re-programming, which is a long job, may result.

Most projects of this kind are one-off and it is essential to carry out a complete simulation of the system beforehand to ensure that the programming of the robots, the weld sets and the handling is compatible and fully integrated.

Control system

Apart from ease of operation, the control system must be straightforward and cost effective to maintain.

At Land Rover, the control panel pin-points the exact stage that has been reached during the operation. For example, the first operation has 35 stages and knowing the exact stage at which any stoppage has occurred ensures that any problem can be located quickly.

The initial control boards used on the system proved to be unreliable and expensive to replace. Knowledge of the reasons for the faults was gained only with experience over a period of time. These resulted from such diverse problems as spikes from the arc weld sets and even from a nearby high voltage electric railway line when a train passed. The boards have been modified since; but basically it was a case of including, for example, more capacitance than was expected from the intial tests.

Weld process

Although tests were carried out to ensure that all required automated welds on the chassis were possible, vertical welds proved in fact a problem in practice. It is the age-old problem of transferring technology from the laboratory to the production environment. The original gas shroud nozzle has been changed in shape and size and robots and jigs have been re-positioned. This helped to overcome welds with difficult access.

However, to maintain the high quality required in the chassis, not all the originally projected welds have been achieved using the robots, although the number has risen drastically since the start. To ensure production flow through the robot facility, these welds are carried out manually.

As mentioned previously, it is essential to overcome weld start problems which become a major concern if, as in the case of the chassis, many welds are involved. Apart from positioning, the material has to be clean. Oil can prevent weld striking at the start.

Bad welds can result from several problems including wire jamming, burn-back, snagging, bad-fit components, and so on. But other causes have been experienced also at Land Rover, for example a draught through the factory can take the welding gas away from the weld. This type of problem is not found under laboratory conditions.

In a production line environment, there can be several components in the process at any one time. This means that if bad welds are occurring, they do not show until the completed component is unloaded and checked, by which time there may be several components with the same fault before the problem, all requiring rectification, is identified. For this reason, it is important to seek some form of on-line non-destructive testing (NDT) methods, enabling any deviation from good welds to be detected immediately.

The NDT would remove also the need for destructive testing of samples, which is expensive both in terms of material and time.

The quality of welds obtained from the robots is more consistent than can be achieved by manual methods and for this reason it is not necessarily appropriate to use the same standards. The Scottish School of Non-Destructive Testing, based at Paisley, have carried out work with Land Rover to determine revised standards for automated welds on the chassis. This exercise ensured that not only the high quality of the product was maintained, but the weld process was made directly applicable to the advanced techniques being used.

Robots and jigs

The selection of type of robot for this sort of system depends on normal criteria such as the number of axes, movement, accuracy, durability, speed, working envelope, and so on. However, the fact that the majority of the operations take place in an unmanned environment, which is normally at a lower temperature, means that the temperature variation also has to be considered.

A problem was experienced at Land Rover, with the robots operating at high speed. The environmental temperature dropped and this caused the oil in the gearbox of the robot to thicken, resulting in a slowing down and ultimately the system going into an emergency stop.

If robots are not specified to operate under various conditions, they may have to be slowed down, resulting in reduced efficiency.

The cabling of the robots can cause problems also. The more flexible the robot, the more careful one has to be in the position and location of the cables for the welding head. If cables prevent free movement, the welding head could be restrained from the weld position. For a similar reason, various locations for the weld wire feed were tried before finding the most suitable.

For any automated welding, the jigs must be much more substantial than for manual welding to ensure that weld position is maintained.

Installation/Area

A robot installation, similar to the type at Land Rover, can take up a great deal of room. Due to the necessary safety requirements enforced, there must be plenty of clearance from personnel around the robot. Also, in the case of smaller installations, there must be sufficient loading stations to gain maximum robot utilisation.

When welding complex components such as the chassis using robots, a certain amount of rectification area will be necessary due to problems in the welding process itself, such as welding gas, wire, and so on. The space required for manual welding operations will be reduced with the experience gained in implementing and operating the system.

At Land Rover, a manual weld facility is available. It is essential for this type of system to include a contingency set-up, for if any part of the system requires maintenance then the whole system has to stop. It is not sufficient simply to have insurance cover for a breakdown as this does not satisfy the product customer, which, in the case of Land Rover, is the vehicle purchaser.

All stations in a system, including those for load and unload, must be balanced to gain maximum system efficiency. All electrical supplies must come from one source to ensure correct phase and must be secure to prevent switching-off.

The Land Rover system is surrounded by solid barriers which form part of the safety precautions. In hind-sight, it would be better to have some form of visual capability from outside in order to see the system in operation, given that operators's eyes are protected from the arc.

When installing a system such as this, the level of safety must be high. It is not sufficient to fence off the installation; it must incorporate such items as pull bars and electronic eyes.

Back-up spares

It is important that adequate spares are available for both robots and weld sets. At Land Rover, it was discovered soon that robots and weld sets, although being of the same specification, differed slightly; so this made interchange difficult. This meant that this method of solving a breakdown was not as efficient as first thought. However, on many occasions it was found to be quicker to replace and re-program the equipment rather than carry out maintenance on the existing machines.

It is essential to have back-up control panels and programming tapes which are continually up-dated to include any modifications to the product. Although these tapes usually need fine adjustment when loaded, they save considerable time in the event of a breakdown.

To save the time taken to re-load the programs from tapes, it is essential to have program storage capability in the event of power loss. This back-up has to be for sufficient time to overcome the power loss problem. At Land Rover, the initial system was for half an hour but this has since been extended to four hours.

Continuing development work

Although much has been achieved already with the Land Rover system, it is an ever-advancing technology, and development work continues to take place to ensure maximum efficiency from the equipment.

To cater for many of these developments, a single robot cell was built. This consisted of a robot mounted on a linear track and servicing two independent welding stations; movement of the robot between stations being DC motor and chain. A schematic layout is shown in Fig. 3.

The weld tables were designed to accommodate various jigs and allow for re-orientation as necessary tenons and tenon slots.

A programmable logic controller (PLC) controls movements, activates locking, monitors sensors and operator controls and selects weld programs.

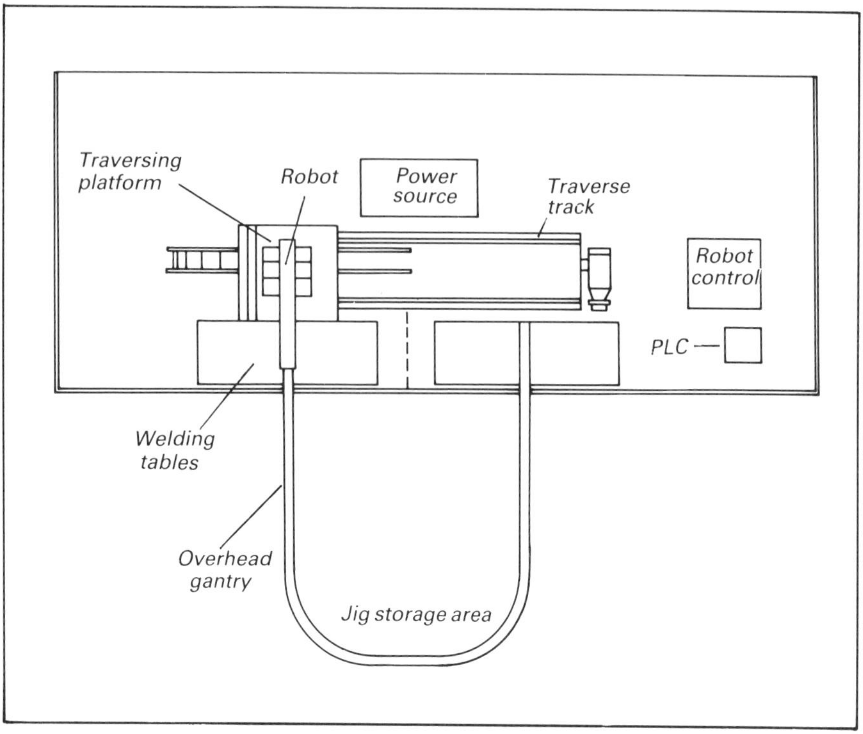

Fig. 3 Development welding cell

This robot cell is now used for continuing development projects. It is used part-time for production when it provides valuable experience for operators and requirements of further production cells.

The main objectives of the overall project, apart from to raise the level of technology of the system, are:

- To ensure that high product quality is maintained by using the latest techniques not only on the main production line but for the welding subassemblies.
- To assess the viability of a seam-tracking vision system. Trials to date have indicated that this technology may overcome many of the problems on tolerances and fit-up as well as overcome the deterioration of the robot accuracy with which a 'blind' system cannot cope. However, although it is generally accepted that vision systems will be beneficial, they may result in a slackening of tolerances which will not be acceptable from a quality view-point. At present, variations in tolerances are quickly found but vision systems may prevent this, causing reject assemblies. Care will be necessary to ensure these problems do not occur.
- To develop a welding quality monitor suitable for the application. This will become more important as weld standards become more stringent.

- To evaluate the feasibility of automating small batch, high variety welding and determine the installation requirements for automating more of the welding at Land Rover.
- To enable new components to be tested for robot welding feasibility.
- To provide a fully operational teaching unit for training of personnel.

Concluding remarks

From the considerable experience gained by Land Rover, much of which has been outlined in the previous sections, the following points can be summarised:

- Avoid an over-complicated system where problems can become more difficult to solve.
- Use computer simulation at the design stage.
- Include as much flexibility as possible; alterations can be very expensive. Make the robots or jigs moveable.
- Use the latest technology but ensure the equipment is fully tested and not experimental.
- Ensure there is an adequate installation area and that the environment is suitable for the application.
- Design jigs for minimum loading time, possibly using robots, to gain maximum utilisation of welding robots.
- Design the system to minimise loss of output in the event of breakdown and ensure there are adequate contingency plans and good maintenance back-up.
- Make sure that component quality from the press-shops is adequate.
- Handling and storage are important. It is not sufficient to have good tolerances on pressings which later are handled and stored badly.
- Ensure components have been designed for automated welding.
- There must be sound and thorough training for operators and maintenance personnel.
- Personnel involved must be those who know what they are doing, not just think they know. It is the people who are working with the job who make it work and solve most of the problems not the people with all the jargon.

The chassis robot welding line has provided Land Rover with invaluable general experience of low-volume, high-complexity robot welding techniques and provided meaningful attention to manual operations.

The future does not pose the question 'whether' robot welding should be used, but 'when?' and 'for what?'. Provided a systematic appraisal is undertaken to assess the complete operation, any user can go forward with confidence of achieving high and consistent quality levels.

Acknowledgements

The author would like to thank Land Rover Ltd for permission to present this paper, Mr D. Bowen, production manager, and Mr F. Townsend, maintenance superintendent, both of the Garrison Street factory where the robot line was installed.

TAURO SYSTEM

L. Bisiach
Bisiach & Carru SpA, Italy

Up to the present moment respotting lines for car and van bodies has tended simply to imitate the old manual lines where the work done per man was restricted to a relatively simple pattern of welds. Having welded, the worker could relax whilst the body was transferred down the line to the next station. This pattern of work, requiring relatively few spots and a rest period (or transfer period), created a situation where the worker was only working for 50–70% of the time. To achieve more than this was not possible, simply because it was inconceivable to ask human operatives to perform to harder work patterns. Creators of the now familiar robot welding lines have copied the architecture of the old manual welding lines and whereas the capital equipment is not so high in the past, it is illogical and bad business practice to work the robots for only 60–70% of the time. This paper describes an alternative to this concept of the line of robots simply mimicking the action of the manual welders – the 'Tauro system'.

The Tauro system (Fig. 1) is a universal system which has been devised to carry a wide range of different types of robot. The system can be up to 40m in length - or longer. It is available in two versions - the portal or gantry machine, and the linear machine.

Options include programmed longitudinal and lateral overhead transfer devices for the automatic handling of workpieces during the cycle. When combined with Bisiach & Carru robots, which have been specifically designed for spot welding applications, the Tauro system is an extremely important innovation for the solution of all spot welding problems involving very large workpieces and low, very diversified production volumes.

The Bisiach & Carru robots have rotating transformers, three-axis hollow heads, which allow cables to pass through them, and enough capacity to accept double-headed guns.

The use of ordinary robots without an automatic gun-changer has meant that it was necessary to have as many robots as there were different types of guns for welding the various sections. The result was

Fig. 1 The Tauro system

an under-utilised robot line and heavy investments devoid of any technological justification.

The Tauro system on the other hand offers truly integral automation for very low volume production. Buses, custom-built cars, trucks, tractors, lorry cabs and similar products can be handled by this completely versatile system.

Tauro system

The system can measure up to 40m in length, according to the customer's requirement and at present it is available in two versions: portal and linear. The main components of the system are: one or two tracks mounted at floor level; one or more mobile robots, according to requirements; two or more welding gun magazines mounted on the carriers; and one or more automatic gun change systems (one on each robot or on the floor).

Other options include programmable longitudinal and lateral overhead transfer devices for the automatic handling of workpieces during the workcycle.

With the Tauro system, thanks to the complete flexibility provided by the interchangeable welding guns, several shorter and simpler lines can be installed to provide a given production at a lower cost.

The Tauro system is a universal system devised to carry a wide range of different types of robot. For example robots can have both single and double guns.

In recent years the idea of using special-purpose machines has become obsolete. This has led to far-reaching changes in automatic welding methods. The automotive industry has devised several original solutions, ranging from rigid transfer lines to robotised systems. All these solutions, however, have one feature in common - it is always the workpiece that has to travel along the line and place itself in front of an operating unit, such as a robot.

Handling also has been a subject of change in itself. Conventional in-line systems have been replaced by wire-guided trolleys. Here again, in most cases it is the workpiece which is moving.

With high production volumes, the idea of using several robots, each of which does a part of the overall work on components moving in front of them, is generally sound, accepted and applied. This idea on the other hand is totally out of place when applied to manufacturing situations involving the low and very low production of a wide variety of products. Too many robots are needed, one for each type of gun needed to weld the given sections. They are in consequence employed at well below their real capacity.

Bisiach & Carru has attacked this apparently intractable problem and devised a tailor-made system for very low production volumes - the only rational alternative being to weld by hand.

The production rate is typically 10 or even 20 times slower than that for mass production situations. However, the universal system requires an automatic change of welding guns; without it the concept would be meaningless.

It would be uneconomic if the line were a closed island and hence a permanent loss of available factory space, as opposed to an area like any other where a robot could be installed as and when required for automatic manufacturing.

The Tauro system embraces two concepts: versatility, derived from the automatic gun changing system with an on-board magazine; and practicality. In this latter aspect both the operators and the equipment used for transport and lifting can move around with the greatest of ease in any part of the area where the system is installed. There are no restrictions imposed by obstacles above ground.

The Tauro system is a longitudinal robot carrying system. It comprises all the services needed for the independent operation of a movable work island. Its main components are:

- One or two tracks below floor level.
- One or more mobile robot carriers.
- Two or more welding gun magazines mounted on the carriers.
- One or more automatic gun changers systems.

There are two versions of the Tauro system: the gantry and the linear. The only major difference between them is the configuration of their movable structures.

The gantry version consists of a gantry track that can carry one, two or several robots mounted both laterally and vertically. It is particularly suitable for the manufacture of large products, such as buses, tractor and truck cabs, and special vehicles such as military vehicles.

The linear version, on the other hand, consists of one or two smaller floor-mounted shuttles, each with a single robot. This arrangement is used when the assemblies and subassemblies to be welded are relatively small.

Tracks

The length of the Tauro system is determined by the customer's specific requirements. The tracks are built in 1m long modules. These can be put together to form lengths that can exceed the 40m usually taken as the maximum. The tracks have been designed to require virtually no foundations. They do not project above floor level. The only sign of their presence is a slot a few centimetres wide. This means that the whole of the area occupied by the tracks is free from danger and fully accessible to workshop vehicles, equipment and staff. The space available is thus put to the most beneficial use.

Carriers

Since the drive units responsible for the programmed movement of the robot carrying structures are mounted on the structure itself, the tracks can be used by a number of structures to form several workstations, each fitted with the appropriate safety devices. In the event of production being stepped up, the Tauro system can be simply expanded to meet the new requirements. The investment will be much lower than that laid out at the start, and there will be no need for extra foundations as the same tracks will be employed.

Magazines and gun change

Each mobile robot carries its own gun magazine. Since welding ancillary operations other than spot welding (such as grinding) CO^2 welding of particular sections, and riveting, may be included in the same operating cycle; the magazine can be used also to hold other units instead of welding guns and so do away with costly changeover operations.

The head of each robot carries an exclusive gun changing system, designed and patented by Bisiach & Carru. It embodies pressure-mounted attachments and assemblies and allows the fluid connectors, current contacts, locators and automatic locking devices to be actuated by means of a single movement.

Automatic loading and unloading of the workpieces can also be made part of the system by the inclusion of appropriately programmed longitudinal and transverse overhead transfer devices.

Application

The Tauro system is suitable for a wide range of applications. Some cases in which it is particularly appropriate include:

- During the start-up of a new car model, when the main production lines still need to be prepared.
- The phasing out of a given product; that is, when the production lines have to be dismantled and it would be too expensive to keep them in operation when output levels fall below the break-even value.
- For sports cars, custom-built models, off-road vehicles, camping vehicles, light trucks, ambulances, trucks, tractors and buses.
- Other possible applications include: electrical appliance manufacture, wire-based products and structures, metal enclosures for the office equipment and electronics industries, and all metal components made in small volumes with a high degree of variety.

How does the Tauro system work out in practice? In the case of a 'conventional' robot line, geared to produce 45 bodies an hour from a single line, the line utilisation is 62.5%, assuming a cycle time of 80s and a transfer time of 30s. In the case of the Tauro system designed to handle the same output, three lines will allow a cycle time per line of 240s. Assuming a transfer time of 30s this will yield a robot utilisation of 87.5%. Simple arithmetic shows that the robots in the Tauro system actually work 40% more than those in the conventional line.

In practical terms 30 robot lines could be replaced by three Tauro lines, each equipped with six robots only, though possibly at least two of the robots would be equipped with double welding guns.

The advantages of the three simple lines are:

- The lines are shorter, more simple and less liable to suffer breakdowns.
- In the case of a breakdown, only one-third of the line is lost - not the complete production system.
- When new models are to be introduced they can be phased-in easily, one line at a time, as the market demands.
- Because the lines are more simple, with fewer machine elements, the changeover itself is simplified.
- Where there are numerous variants in the product range it is easier to organise the production using three lines.
- The Tauro system enables the investment to be phased so that it is aligned to the reality of the market situation. It avoids over-investment in equipment which may never be used to capacity, if the original estimates of demand prove to be over-optimistic. Conversely, if the demand exceeds planned production capacity the problems can be very grave in the case of conventional production lines, as the cost of a second line is likely to be prohibitive and the space requirements excessive.

The Tauro system therefore clearly offers numerous benefits from the financial standpoint.

Tauro as a standby robot

A Tauro gantry robot can be used to great effect as a standby robot at the end of almost any robot welding line, regardless of the robot line manufacturer.

It can be equipped with a magazine of several guns to enable it to take over the work of any of the robots up-stream and which may break down or may be in need of a major overhaul. In the case of a breakdown, current practice is such that incompletely welded items pass to a manual station to be finished off. This reduces the level of quality, greatly reduces the production rate, causes organisational inconvenience, and a loss of product leaving the factory gates.

The Tauro robot would be brought into operation automatically whenever a preceding robot breaks down, since it would have the appropriate welding gun in its magazine and the relevant welding program in its memory. It could undertake a wide variety of work - effectively undertaking the work of any robot on the line. Also, as robot welding lines become older, serious maintenance work is required which may mean taking the robots out of line for several days or even weeks.

As there are only a limited number of days available in the year it is very difficult to maintain the robots effectively. With a Tauro robot at the end of the line, equipped to undertake any of the work of the up-stream robots, a robot can be removed from the line for stripping down or rebuilding; it could even be possible to return it to the manufacturer for a major overhaul and possibly even for updating.

In addition, the Tauro robot could carry programs enabling it to undertake work required by special models which are sent down only very occasionally.

FLEXIBLE AUTOMATED BODY ASSEMBLY

G. Repetto
Fata New Hunter Engineering SpA, Italy

The use of robots for welding has made significant progress in the automation of body-in-white manufacture. The next major change has been to allow multiple body styles to be assembled automatically on the same line. This allows not only complete flexibility of production but cuts dramatically the cost of model changes or new models. This paper looks at the techniques which have been adopted in several systems to achieve this and takes a look at some of the options now available.

In recent years, insistent calls for product modifications and an ever fiercer competition in the market have meant that the principal automakers have been faced with completely new problems, namely:

- The need to produce several different models at the same time.
- The need to ensure the extension amplification and/or modification of their current production range.
- The need to be able to carry out a quick, simple, but radical change in production.

All of these need to be satisfied through the search for:

- High productivity.
- High reliability.
- Simple plant management and maintenance.
- Maximum payback on investments.

FATA's contribution

A typical example of the results of research can be seen in the latest welding lines designed and built by FATA for the plants of the same European automotive group.

These lines are virtually the same in all three plants. They have a capacity of 80 body shells per hour. A total of about 580 spot welds are applied to each shell.

In keeping with the technological criteria based on which the car has been designed, these lines extend in single lines from reception of the complete floor through the:

- Chassis number embossing.
- Bodyside loading and toy-tabbing stations.

Then, as far as the final respot welding lines, they become double lines through the:

- Bodyside framing.
- Roof loading.
- Roof welding stations.

FATA-Gate

The most important station and the true heart of the installation is the bodyside framing station, known as the FATA-Gate. Here the vehicle receives its final geometry through the work of special equipment known as side-gates. There is one for each side. Then there are robots that apply about 90 spot welds.

The station consists of an electro-welded bearing structure resting on a base adjustable for levelling, and a series of devices (primary and safety) to direct and control the movements whereby the side-gates are interchanged.

The FATA-Gate can accept three different pairs of side-gates carrying the locations and clamps for three different vehicle models. A further increase in the versatility of the framing stations can be obtained by equipping it with several types of side-gates, since the side-gates are fully interchangeable, and by providing it with an overhead buffer.

The arrangement adopted by FATA to handle the three pairs of side-gates is based on a verrtical movement along their longitudinal axis where there is less encumbrance. The result is that the FATA-Gate occupies only one step in the technological line and has reduced overall dimensions.

It is important to note that this station is based on the use of modular components, giving rise to a standard project adaptable to a very wide range of production requirements (cars, commercial vehicles, minibuses, etc). This means that the station possesses the highest degree of versatility, even for the acceptance of future models, whether large or small.

The handling system employed to feed and remove the bodies within the station is of the 'lift & shift' type, using skids only as handling means. In the FATA-Gate, as in all the other technological stations that make up the FATA welding line, clamping of the body in the working position is brought about by model-specific devices. These release the body from the skid during its transit and assign it the correct location in space.

The working cycle of the FATA-Gate is as follows: each skid has a magnetic plate with the model code; this enables the fixtures on the technological stations to be preset. In other words, the framing station knows in advance the type of vehicle in transit, and hence the right type of side-gate to use.

If the pair of side-gates is the same as that already in the working position, the station opens their locking devices and shifts them sideways just enough to allow the finished body to be removed. It then brings them back to the work position and locks them as soon as the incoming body is in position.

If, on the other hand, the side-gates have been changed, they are removed vertically as already described. Since these movements are related to those of the lift and shift system, it follows that the idle time provoked by the changeover is virtually nil; the only thing required is the bringing up of the clamps to the body when it is in position on the station locations.

With the fixtures in this position, the automatic tack-welding operation begins. This is performed by the six or more robots manning the station. After this operation, the stages previously described are repeated to release the finished body and make preparations for the entry of the next body.

The working cycle is managed automatically by two programmable controllers (PLC), one of which is dedicated to the drives and to the control of the on-board side-gate clamps, while the other manages the dialogue with the welding robots and the movements in the lift and shift line.

These two PLC's, like all the other PLC's in the different zones of the welding line, converse with the higher level, namely the central system. This is responsible for all the control, management and diagnostic operations of the line as a whole. In its turn, it is interfaced with the computer that controls the entire workshop.

Looking to the future

After originating what can truly be called 'up-to-date technology', FATA's research sector is continuing its efforts with an eye on the future in providing more and more flexible plants.

This future can be glimpsed in FATA's extension of the use of automated guided vehicles (AGV) to bodywork welding.

After building the first installations, among which is a significant door welding system for the Soviet Union, in which AGVs were paired with the welding robots to reach a cycle-time of one door every 18 seconds. AGV installations are currently being commissioned for a US company for use in several plants as handling systems within the workstations of various departments (for example, body-in-white, trimming, body welding, undercarrier operations, and so on).

BONNET ASSEMBLY CELL AT JAGUAR CARS

T.J. Vickerstaff
Jaguar Cars, UK

The paper describes a flexible cell for the assembly of bonnets (or hoods) using three robots for adhesive application, spot welding and handling operations, respectively. The cell has been supplied by Dainichi Sykes Robotics (DSR) and forms part of the technical collaboration agreement between that company and Jaguar Cars. In addition to reviewing the operation of the cell the safety features incorporated into the cell are explained and reference is made to the Jaguar Cars Safety Standard.

Jaguar Cars manufactures luxury cars, the current range including the XJ6 saloon, the XJS coupe and the Cabriolet and the XJS with a targa top. Until 1984 Jaguar Cars was part of the BL Group but in August 1984 was privatised and is now a completely independent company.

In the late 1970s the performance of the company was not good. By 1980 sales had declined to 13,000 vehicles per year, man-hours per car were very high and the quality of the product was poor. At that stage BL decided to appoint a new chief executive whose brief was essentially to decide whether or not the company could be saved and to make appropriate recommendations.

John Egan, chairman and chief executive, quickly decided there was a real chance of turning Jaguar Cars into a successful company, provided that everyone could be persuaded to that belief. A plan was developed with three key objectives:

- Improve quality and reliability of the product.
- Reduce operating costs through improved efficiency in all areas of the fixed and variable costs, but with particular attention to production.
- Improve communications within the company to involve all levels of employees, to explain the real situation of the business, to outline the plan for survival and to win the hearts and minds of everyone at Jaguar.

The success of the approach is evidenced by the results achieved in 1984. Compared with 1980 sales worldwide have increased by 150%, productivity has more than doubled and quality has improved substantially. The latter point is of course one of the major reasons for the dramatic increase in sales. The final high point in 1984 was the successful privatisation of the company.

One of the most important points about these improvements is that they have been achieved without major expenditure in new facilities, automation, and so on. There was no money available for such investments and consequently the results were achieved primarily by doing the same things, but in a more efficient way.

Jaguar now has plans for further expansion and new models. This next phase in the development of the company will require significant investment in new and advanced technology to give further improvements in quality, increases in productivity and reductions in cost.

In the latter half of 1983, by which time it was clear that Jaguar Cars was successful and would need to invest in new technology, it was decided that robotics should form an early focus for such investment. Jaguar Cars had no experience of robots and in order to make progress quickly a technical collaboration agreement was reached with Dainichi Kiko (DK) in Japan and Dainichi Sykes Robotics (DSR) in the UK. DSR are systems engineers using DK robots and in conjunction with Jaguar engineers conducted a survey of Jaguar Cars to identify potential areas of application. By the end of 1985 some 20 robots were installed in various areas including arc and spot welding, polishing of engine ports and assembly operations.

This paper describes a cell for bonnet assembly installed in the Jaguar Cars body-in-white plant and incorporating three DK robots.

Cell details

In automobile terms Jaguar Cars is a relatively low volume producer. Consequently, to justify an automated cell it is often necessary to incorporate significant flexibility so that the cell can accommodate a variety of models. The bonnet assembly cell has been designed to manufacture the bonnet for the new XJ40 model and a further new model scheduled in the company forward plan.

The system (Fig. 1) occupies an area of almost 60m^2 and is capable of producing a finished welded bonnet every 2.5 minutes. It incorporates three DK robots: the FR10, PT1200 and BA3200 types which carry out adhesive application, spot welding and inter-operational handling operations, respectively.

There is a single cell operator who has two loading duties:

- Loading bonnet outer panels to a buck on a linear bed for movement into the adhesive application station.

- Loading bonnet inner and other components onto a three-station carousel for movement into the spot loading station.

A computer controller acts as host for the cell, controlling and monitoring over 500 separate sensors, solenoids and switches.

The system's operational cycle begins when the operator enters the area adjacent to the adhesive buck linear shift table (Fig. 1), and in so doing stands on a pressure pad. The pressure pad isolates the adhesive robot (FR10) and the buck linear shift, preventing any movement. The operator unloads a bonnet outer panel from its pallet and places it, outer skin down, onto the buck, using a simple manipulator. On stepping back from the load station the operator releases the pressure pad override and presses a cycle initiate button. The buck with the outer panel indexes forward into the work envelope of the FR10 robot which operates in the horizontal plane to lay a pattern of adhesive onto the inner surface of the panel.

While adhesive is being applied the operator approaches the carousel portion of the cell (Fig. 1). Once again pressure pads isolate the area involved, i.e. prevent a 'carousel rotate' signal. The operator loads an inner panel and eight related components from a pallet and tote pan racks, respectively, into a spot weld fixture at the load station of the three position carousel.

The final operator loading duty is to step back and press a cycle initiate button causing the loaded carousel station to be indexed into the working envelope of the PT1200 robot where 44 spot welds are carried out. When the welds are completed the carousel indexes again into the third station to await the handling robot.

The BA3200 robot has a number of handling tasks to perform. Firstly, when adhesive has been applied to the bonnet outer the robot transfers the outer to a fixture (Fig. 1) so that the outer can receive the assembled inner. The handling robot then picks the assembled inner from its wait station and places it on top of the outer. The PT1200 spot welding robot then moves on a linear shift from its original position into a new position where it can spot weld the inner to the outer.

After completion of all welding operations the BA3200 robot picks up the assembled bonnet for transfer along a linear shift into the clinch press (two are scheduled) where the clinching operation which completes the bonnet is carried out. Finally the completed bonnet is placed onto an output shuttle by the BA3200 robot and automatically conveyed out of the cell.

Future cell development

At present the cell is effectively an isolated 'island of automation' in that components are fed to it by hand and the completed assembly, after leaving the cell, is moved manually to its next operation. Furthermore, the flexibility inherent in the adhesive and spot welding elements of the cell is not reflected in the clinching area where dedicated tooling is provided, therefore limiting the overall cell flexibility.

Potential future developments to overcome these problems are concentrated in three areas:

- Automated part handling to the cell and possibly into the jigs and fixtures on the adhesive buck and the spot weld carousel.
- Flexible clinching process using either quick-change (in-process time) tooling or an entirely different process such as nibbling with a robot or similar device.
- Automated handling of the completed bonnet into an automated hinge fitting area.

The bonnet cell has been laid out to accommodate such future developments.

Safety aspects

In the early days of Jaguar Cars' involvement in robot-based projects, wide-ranging discussions were held with personnel at all levels to explain the purpose of the Jaguar/DK/DSR agreement and to indicate those areas which would be affected. Not surprisingly, safety was one of the topics which came up for discussion and accordingly it was decided that a Jaguar Cars Safety Standard was required. The Standard has been developed in conjunction with DSR and is now used as the basis for all aspects of cell design as they relate to safety.

Particular reference has been made to BS5304, Safe Guarding of Machinery and to a publication produced by the Machine Tool Trades Association on the Safe Guarding of Industrial Robots.

Enclosures

A typical manufacturing cell such as the bonnet cell has two types of enclosure:

- Control enclosure containing all cell controls.
- Manipulation enclosure containing the robot and related equipment.

The cell has two gates, one providing access to the control enclosure and the second leading off the control enclosure into the manipulation enclosure. The separation of the two enclosures affords the possibility for maintenance and other skilled personnel to work on the controls without concern about the moving elements of the cell. Furthermore, the need to gain access to the manipulation enclosure via two gates serves as a constant reminder that entering the manipulator area is potentially dangerous unless adequate precautions are taken.

Access

Access for part loading by the operator from outside the enclosure perimeter is through counterbalanced gates built into the guarding so that after loading the gates are manually closed. Closure is detected and a shot bolt is activated to retain the gate in the closed position. The shot bolt passing through the gate enables the cell to be energised. At the end of the productive cycle the shot bolt is released and an unload light appears on the operator pendant. Sensors to detect correct part loading can override the shot bolt and prevent its closure so that an operator can gain access to make an adjustment.

To access a robot for tool changing purposes, such as the tips on the PT1200 spot welding robot, access is available from outside the fence through a gated hole, maximum size 600 × 600mm. The operator can command the robot to move to the tool change station at which point all motive power is removed from the robot and the operator can make whatever change is necessary. Once again a shot bolt mechanism is used for the gate.

If access to the manipulation enclosure is necessary for an operation which does not relate to the robot, for example repositioning a component, wire changing and so on, a special key arrangement is used. The person making the adjustment, who could be the operator, has a key to to the control enclosure. On using that key, a further key is released which fits the main systems control panel and switches off all motive power, putting tooling into manual mode. A further key is now released which opens the manipulation enclosure. On completion of the operation the reverse procedure is adopted.

If it is necessary to move the robot to its 'home' position to gain access to an area normally obscured by the robot, the person doing this - not the operator - has special training and a special key which in addition to releasing the manipulation enclosure from the control panel also energises a robot pendant in the manipulation enclosure. This pendant only allows robot movement at slow speed. On entering the manipulation enclosure the person concerned uses the robot pendant to drive the robot back to its 'home' position.

For robot programming a further variant is employed because the person concerned - normally a senior authorised person - has a special programming pendant which is taken into the cell and used instead of the slow-speed robot pendant mentioned previously. After each program change a spare tape of the new program is prepared and stored.

Pendants

The cell has a master operator pendant. In addition to special features applicable to the particular cell being designed, all master control pendants have the following features:

- 50mm red mushroom retaining emergency stop button on yellow background.
- Green shrouded 'master start' push button.
- Red shrouded 'end of shift' push button.

- For each robot a red unshrouded 'robot hold' push button and a green shrouded 'robot restart' push button.
- Green 'systems home' status indicator.
- Red 'emergency stop' status indicator.
- White 'end of shift' status indicator.
- Green 'cell healthy' status indicator.
- Red 'robot hold' status indicator for each robot.

The cell robot pendant - the pendant inside the manipulation enclosure used for slow speed movement - has the following features:

- Manual robot functions - at slow speed only.
- 15mm emergency stop button.
- Black push button as a 'dead man' control, i.e. pendant only operative when the button is pressed.

The programming pendant is similar to the cell robot pendant but has all the normal robot controls, automatic as well as manual, available.

In some respects the emphasis on safety aspects may seem to be excessive. However, experience has shown even with a single robot system supplier, that problems can occur unless a safety standard is prepared and adhered to whenever possible.

Inevitably there will be occasions when a compromise is necessary. With the bonnet cell for example, the linear shift axes used on the PT1200 spot welding robot and the BA3200 handling robot effectively prevent the use of completely independent manipulation enclosures for each robot. This is not a problem providing that the difficulty is appreciated at an early stage and appropriate precautions taken. The Safety Standard provides the base against which each example can be tested.

Concluding remarks

As with all automation, the Jaguar Cars move into the use of robot-based islands of automation has not been without its difficulties. Component variability which can be absorbed using manual techniques frequently cannot be accommodated in an automated situation. The oft-quoted solution to this problem is to design for automation but in practice, as many will know, this is not always possible and efforts must be made to build adequate flexibility into the facility.

A further problem has been the need to move up to a learning curve in a number of areas. Jaguar Cars' engineers have had to learn that in the majority of installations involving robots, the robots themselves are the least troublesome element. Integration of the various cell elements often leads to more serious problems.

Training has had to accept the need for new programmes with different emphasis. In this respect Jaguar Cars is fortunate that training receives a high priority - indeed the first robot installed at the company was in training. Computer-based modules for teaching principles of robots and robot programming were widely used in open learning courses offered to the workforce.

The bonnet cell is physically the largest robot-based installation at Jaguar Cars. To date the indications are that it will be able to achieve its objectives in terms of cycle time and flexibility and will significantly improve the quality of this part of the Jaguar body.

CAR BODY ASSEMBLY WITH ASEA 3D-VISION

R. Svensson
ASEA Robotics AB, Sweden

Assembly of cars includes many operations where parts like doors, fenders, wheels, windows, and so on, are mounted on the car body. These operations are today performed manually as a result of difficulties in measuring and adjusting gap and flush between part and car body. The use of the ASEA 3D-vision system in such operations makes it possible to localise an object very accurately in space and to measure the relative distance between two objects. Together with the good positioning accuracy of the ASEA robots the result will be car parts assembled with a high degree of precision. Practical examples of applications in the car industry demonstrate the capacity and possibilities of the system.

The new ASEA 3D-vision system is used together with the ASEA range of Irb-robots. It includes all the advantages of the ASEA 2D system, such as the grey-scale gradient concept and the unique user-interface, that makes the system easy to learn, use and program. One application for the system is automatic measurement and adjustment of parts when assembling cars.

As ASEA vision is a grey-scale, gradient-based system, which is capable of processing images of metal parts. The surface texture of these parts can vary due to welding, oil and, for painted parts, different colours.

Assembly of car bodies includes insertion and fitting of many different parts. For some of them a 2D system is sufficient, namely where the distance to the car body is known. When, however, information about the part's position is needed in three dimensions a 3D system is required.

A true 3D system should be able to measure the distance to all pixels in the 2D image and with the same resolution as in the image plane.

Depending on the problems to solve there are a wide variety of 3D systems that differ in how the distance information is obtained.

There are a number of ways of achieving the distance information: optical triangulation, stereo vision, focusing and ultrasonic.

3D principles

Optical triangulation using structured light

The light source (such as laser) is mounted at a precise angle to the optical axis of the camera (Fig. 1). Using trigonometry the distance between two objects can be calculated. This method can be used for both relative and absolute measurements. It is possible to measure the distance to one point in the scene or to measure the distance to a number of points along a line, a grid, and so on.

Stereo vision

The scene is observed from two different positions with two cameras (Fig. 2). The two images are then combined and a 3D image is constructed. This method usually requires considerable computer power and in many cases gives more information than needed.

Focusing

Focusing (Fig. 3) is a method that uses the fact that a sharp picture has higher contrast than a blurred one. A section of the image can be used to determine highest contrast by 'focusing' and the absolute distance to the scene can be calculated. This method can be time-consuming and requires moving mechanical parts.

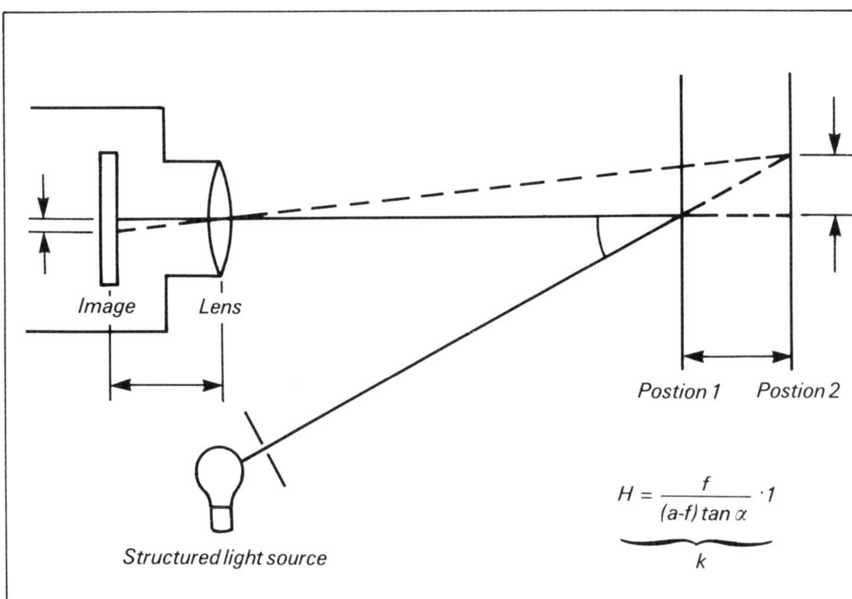

Fig. 1 Optical triangulation

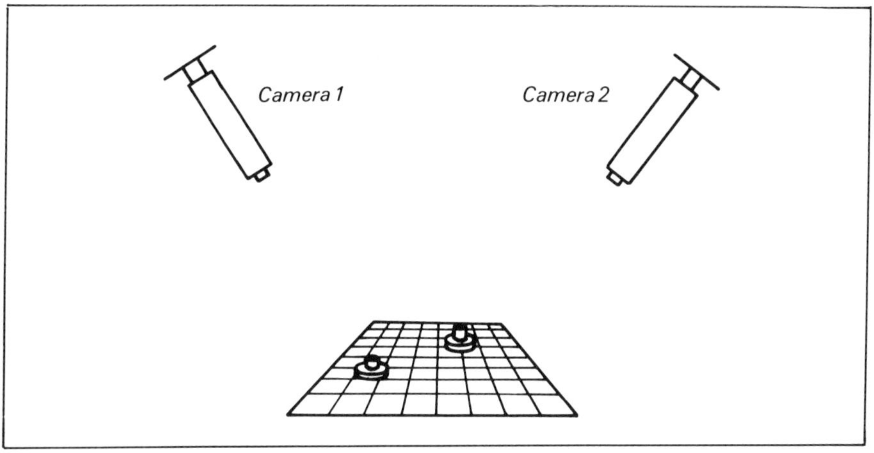

Fig. 2 Stereo vision set-up

Ultrasonics

In recent years different ultrasonic sensors have been developed. These work in air and can be used to achieve distance information. The problem with ultrasonics is the sensitivity to environmental changes, such as air pressure, temperature, and so on. There are two concepts - the pulse system and the continuous transmission frequence modulation system (CTFM) (Fig. 4). The pulse system uses the time between sending a pulse and receiving it as distance information. The CTFM-system uses either the difference in frequency between the transmitted and received signal of the time difference between the points where the outgoing and incoming signals have the same frequency.

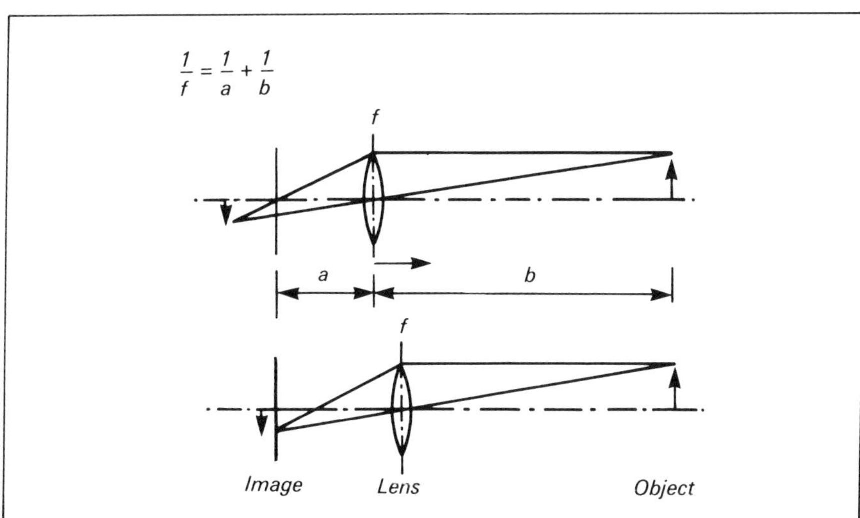

Fig. 3 The principle of focusing

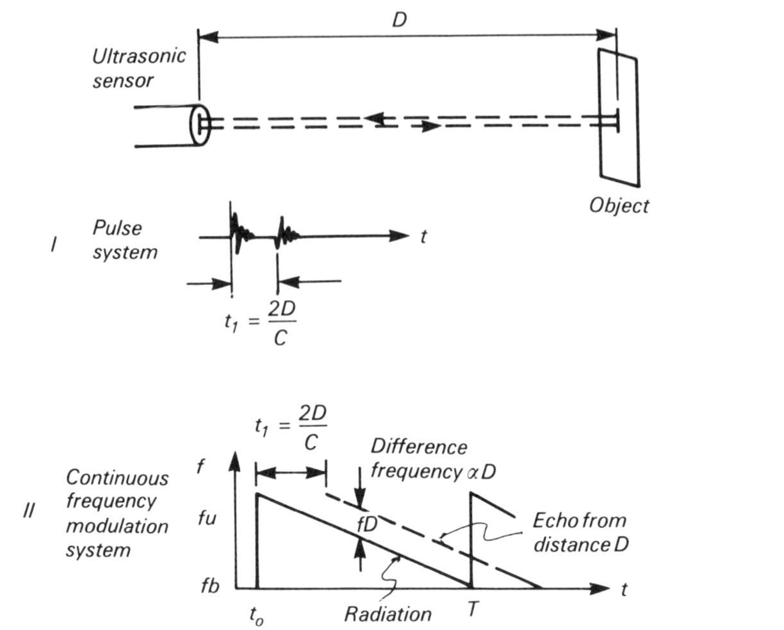

Fig. 4 Ultrasonic sensing

The ASEA vision system

ASEA's new vision system is a three dimensional grey-scale gradient system. The 3D principle used in this system is optical triangulation. Among the requirements for this system are that it should be a general system to cover a wide application range, it should be fully integrated with the robot and it should have a user-friendly, efficient man-machine communication for 'non-software' people.

As the earlier generations of ASEA's vision system only had the capacity to identify closed contours (domains), the new system has been improved to handle the identification of segmented contours. This feature is used to find the projected lines on the car bodies.

The use of an ASEA-developed image processor, implemented in hardware, gives sufficient speed for the normally very time-consuming 3D image processing. The different surfaces of metal parts make the software-controlled contrast and brightness values used in the image processor very useful. Images from up to four different cameras are combined to calculate the position where the object should be inserted. These calculations include transformations for six axes simultaneously.

Insertion

Insertion problems can be divided into two separate cases. The first case is when the object can be roughly inserted into the opening before the measurement and then adjusted relative to the opening. In this case the camera can see both the object and the edges of the opening. The gap is measured and adjusted to a certain value. These measurements are relative.

The second case is when the system first has to locate the opening and then insert the object. That is, the measurements are absolute. In most applications of this kind the required accuracy makes it necessary to use more than one camera to locate the opening where the part is to be inserted. To find the plane of the opening it is also necessary to measure the distance in three to four points on the edge of the opening.

When all coordinates for the three or four points have been measured the system should calculate a robot movement to insert the part.

Measurement principles

Relative measurements

Before the measurements are taken the object has to be roughly positioned in the opening. The gap is measured in the scene plane (*X*- and *Y*-coordinates) and then the relative distance between the object and the edge of the opening (the 'step' in the *Z*-direction) is measured (Fig. 5). This is done in each point.

Absolute measurements

The gripper with cameras is positioned to measure a position just outside the opening. Then the absolute coordinates (*X, Y, Z,*) for three or four points on the edge of the opening are measured (Fig. 6).

Calculations

The measured differences in the three points together with the true values and tolerances for the gaps are then used to calculate a movement vector in each point. These vectors are used to calculate a translation and rotation of the robot hand, in six coordinates, to insert the object in the opening with the required tolerances.

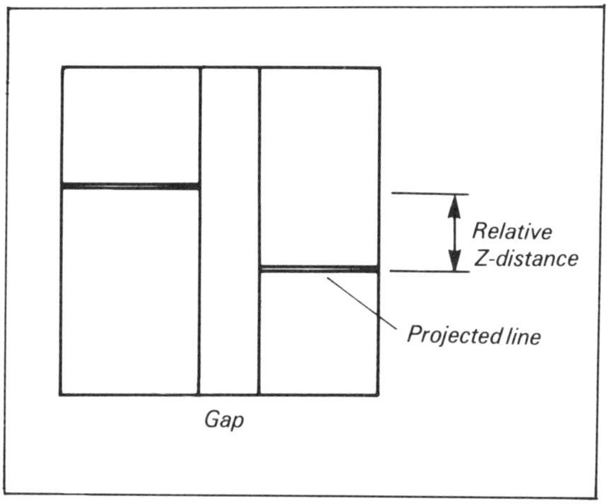

Fig. 5 Image from one camera when a relative measurement is done

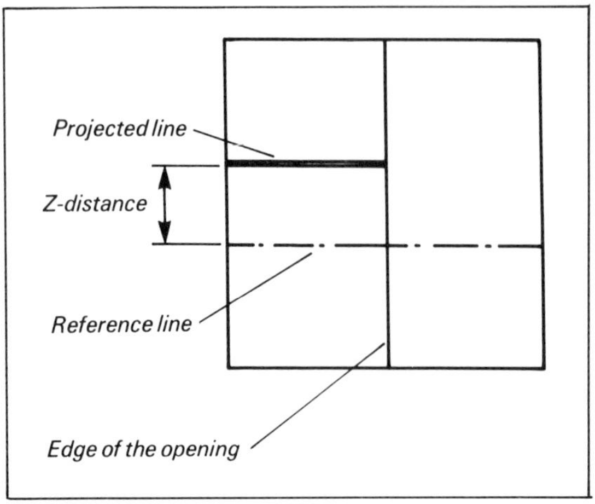

Fig. 6 Image from one camera when an absolute measurement is done

Programming

The program can be divided into three parts: calibration, definition of measurements, and robot programming.

Calibration

Calibration is usually done once per installation and includes adjustment of cameras, finding scale factors and transformation matrices between the camera coordination systems and the robot coordinate system.

The following steps are performed for each camera:

- *Scale factors.* A picture with three equally distanced circles is placed in the scene. The distance is manually fed into the system. An image processing is ordered and the scale factors are then automatically calculated.
- *Transformation matrices.* A pin with three fingers is placed in the robot hand. The robot is manually positioned to get the pin in parallel with the normal to the scene plane. A dark circle is placed where the pin points at the scene and the robot position is stored. The robot then moves back to the position where the camera can see the scene. The robot automatically moves with the camera in parallel with the scene while images are being processed. With the obtained data, the scenes orientated in space and the transformation matrices to robot coordinates are calculated. This is done for each camera.
- *Rangefinder.* The rangefinder is factory-calibrated to the required measuring range. The range is manually fed into the system and the scale factor for the distance measurement is calculated. The robot hand is placed in the measure position with the car body in its optimal position and by image processing, the reference line for the distance

measurements is found. The calibration can also be performed in a calibration fixture.

Definition

The definition procedure includes definition of models and measurement sequences. A model defines the features that should be used to identify and localise the measurement point in the image.

A measurement sequence defines what kind of measurement is to be done. It includes a sequence of measure points, each point defining the following: camera number; model number, used for identification; true value for measurement; and tolerances for the true value. The measuring point also includes statistical data for the measurements.

Robot programming

The robot's instruction set includes the instruction 'MEASURE', which initiates a measurement according to the selected measurement sequence.

Applications

The interest in using vision in final assembly of cars is high. Parts that benefit from being assembled with vision include: wheels, windshields, hoods, doors and fenders. In an assembled operation of a hood, for example, it is necessary to measure both gap and relative distance between the hood and the car body (flush). Here a 3D system is required.

To reduce the need for accurate positioning of the car body and to obtain a flexible system for this kind of assembly operation, a vision system with the cameras mounted in the robot arm offers a solution.

One application, where the ASEA 3D vision system is used, is adjustment of gap and flush when doors are mounted on car bodies. Another is automatic insertion of windshields. The door adjustment includes relative 3D measurements and the windshield insertion absolute 3D measurements.

Door adjustment

In this case (Fig. 7) the car body is positioned accurately enough to allow the robot to roughly insert the door in the opening before measurements are taken. The measurements are then taken relative to the door and body. If the car body is not positioned accurately enough, an absolute measurement has to be done before the door is inserted.

The gripper is a flexible tool designed to grip and handle both the rear and front door at the assembly operation. It is equipped with three vision sensors to measure the gap and flush at three points along the edge of the door opening. The vision sensors project a 'half-moon' on the scenes for triangulation measurements of the distance. The gripper is also equipped with a screwdriving unit for tightening of bolts.

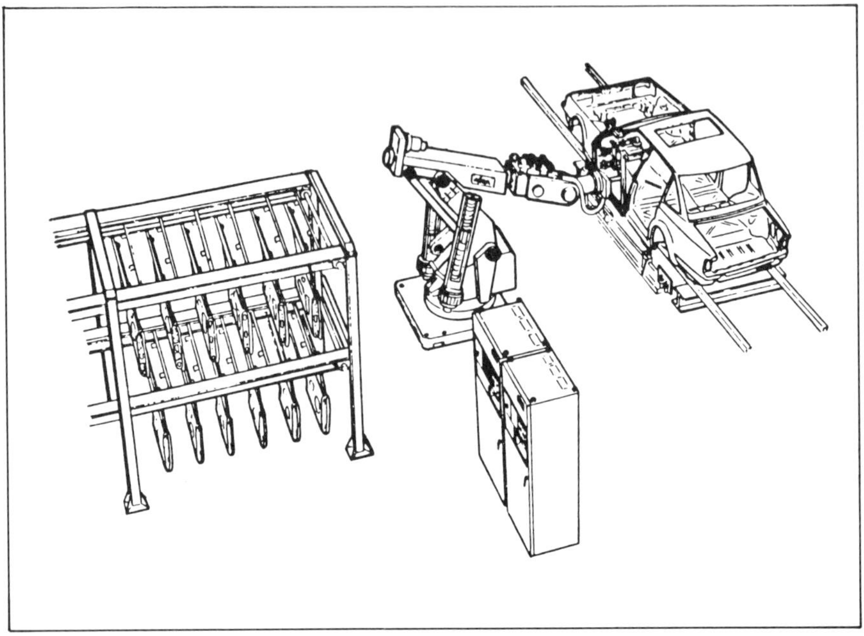

Fig. 7 Door assembly

The robot, with the door in the gripper, moves to a small distance from the body surface. Then the vision system, at three points, measures the relative distance between the car body and the door in three dimensions. From these measurements a new optimised robot position is calculated relating to the true values set by the operator. The robot is then moved to this new optimal point.

In the calculations, allowance is made for the weight of the parts that later will be put in the door. This makes final adjustments on the complete car unnecessary. The position of the door in the gripper may be less accurate since the measurements are always relative to door body opening. The screwdriver unit, placed in an arm on the gripper and controlled as the seventh axis, is positioned through the door window opening and fastens the hinges to the body from the inside of the car. When the door is fixed in place, the robot opens the door slightly, releases its grip and moves the fixture where a front door with hinges has in the meantime been provided. The robot then repeats the same sequence for the front door.

Windshield insertion

The gripper is equipped with four vision sensors to measure four points, all along the edge of the car body opening. The sensor uses an infrared laser to project a line across the scene for triangulation measurements of the distance.

The windshield is held by suction cups on the gripper (Fig. 8). The robot positions itself at a point just outside the windshield opening and

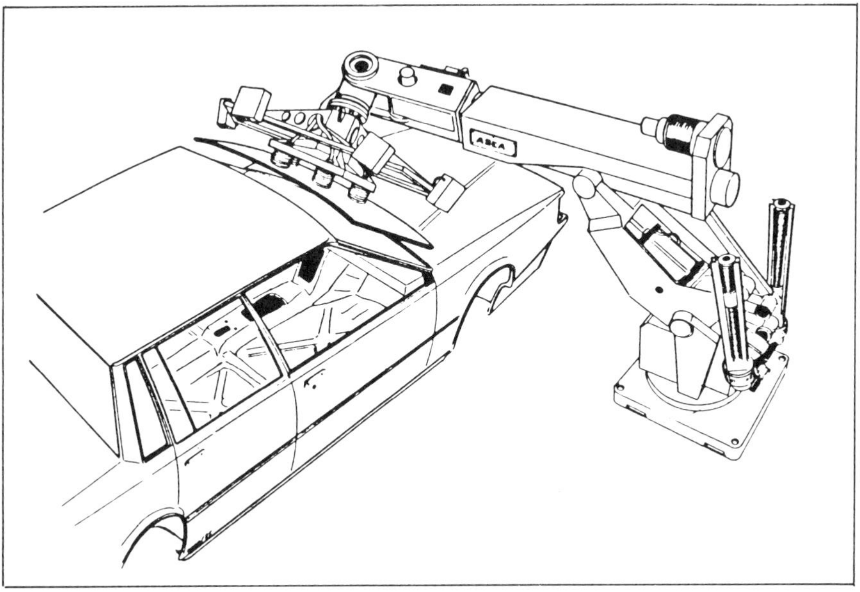

Fig. 8 Windshield insertion

the vision system, localises and measures four points on the edge of the opening. The results of these measurements are the coordinates of all points in three dimensions. When the position of the windshield in the hand is known, a robot movement can be calculated to insert the windshield into the opening.

Concluding remarks

Increased flexibility, simplified design, reduced design and installation costs, and the possibility of standardising the supply system in final assembly are advantages obtained when a vision system is used together with a robot.

AUTOMATIC BRAZING OF CAR BODIES BY INTELLIGENT ROBOTS

T. Kudo and M. Suzuki
Nissan Motor Co. Ltd, Japan

The arc brazing of car body joints is usually carried out manually since it is a difficult process to automate. Difficulty in automatic arc brazing arises from the fact that the joint position varies due to the deviation of body shapes and the positioning of car bodies on the conveyor. To solve this problem, Nissan has developed an intelligent robot system that finds, memorises and tracks the joint line automatically. This system is now in operation on the body assembly line in the Nissan Tochigi Factory.

Until recently, the brazing of car bodies has depended on manual labour, since conventional playback robots are not suitable for the task. Their unsuitability is related to the fact that the positions of brazing lines vary from body to body, due to the inherent spring-back of car body parts, variations in the mounting position of parts on carts, and conveyor positioning inaccuracy. Furthermore, improvement of positioning accuracy to a level which would permit the use of playback robots has been found to require prohibitively large investments.

To obviate these problems, Nissan developed an intelligent brazing robot system for use in automatic plants, capable of offsetting the brazing path according to the deviation detected by a position detector.

There are several kinds of joint position detectors. Optical sensors are used in some detectors – some of them recognise the joint position by analysing the slit beam image projected on an object [1]. Some find the joint position by measuring the distance to an object using a point of light [2].

Experience in robot vision development shows that optical sensors are not only complicated and expensive, but their operation may be adversely affected by certain light sources and differences in the shape, colour and surface condition of the object to be recognised, making them unreliable. Among other kinds of detectors are arc sensors that detect joint positions by monitoring arc voltage. However, they cannot be applied to thin panels.

Consequently, Nissan developed a contact-type, lap-joint sensor which realises a highly reliable automatic brazing line at low cost. The new brazing line is in full operation at the plant.

The robot system

Components

Fig. 1 gives an outline of the system, which comprises the following components:

- Robot – a six-axis controlled, multi-jointed robot which has functions to recognise and record its current coordinate values, using an external trigger signal.

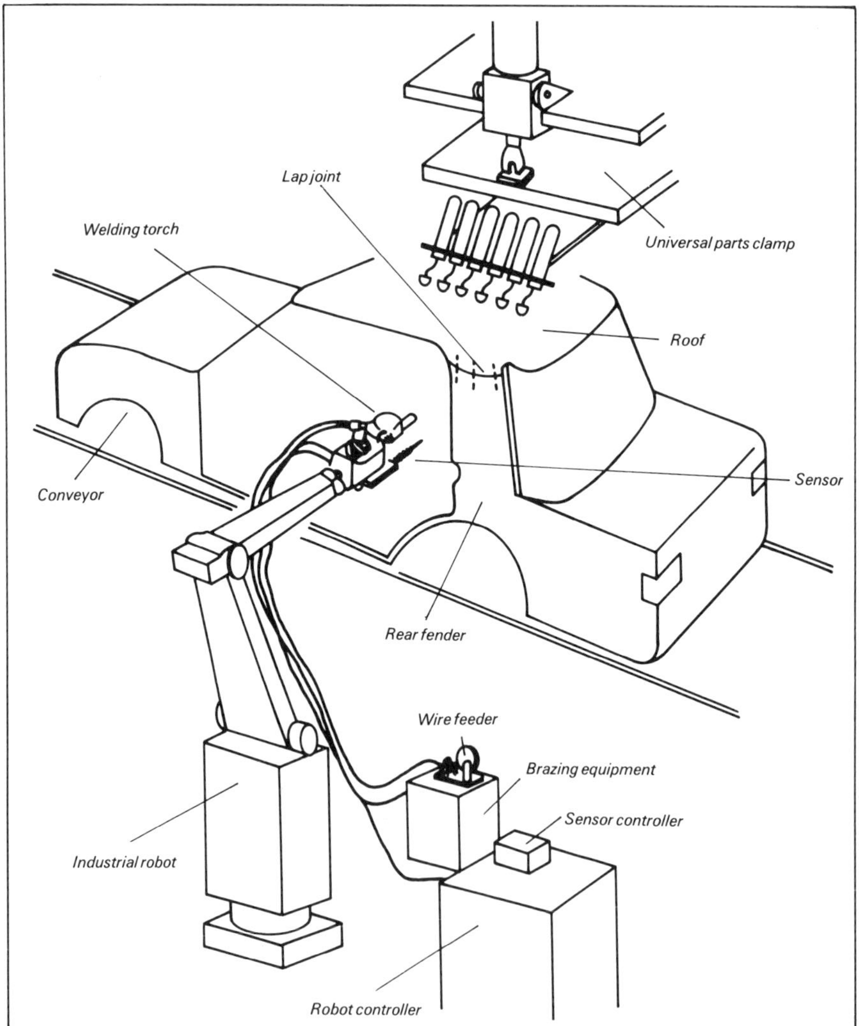

Fig. 1 Automatic brazing system

- Lap-joint sensor systems - this incorporates a linear transducer. The sensor controller continuously monitors the sensor output. When a lap joint is recognised, the sensor controller outputs a detection signal to the robot.
- Universal pad - holds the joggle near the lap joint, maintaining an optimum brazing gap to ensure brazing quality.
- Brazing equipment - plasma arc brazing equipment and its accessories.
- Line controller - processes car model identification signals and conveyor interlocking signals.

System operation sequence

When the robot controller receives a car model identification signal and start signal from the line controller, it moves the universal pad to a specified position according to the car model. The robot controller sends a command to the pad controller so that the universal pad holds the lap joint. The robot places the tip of the sensor on the car body parts, then moves the sensor to where the lap joint is supposed to be. When the sensor controller detects the lap joint, the controller sends a joint detection signal. When the robot receives the detection signal, it records its current coordinate values, stops probing, and retracts the sensor. After detecting a sufficient number of points for brazing, a brazing path is calculated. The robot then brazes the car body according to this path.

After brazing is complete, the robot retracts the universal pad and sends a cycle completion signal to the line controller.

Sensing system

Recently, optical sensors have been used as joint position detectors, but their performance is not so reliable, since certain light sources, colour or shape of an object affect the sensor output.

To overcome these disadvantages, Nissan developed a lap-joint sensor, details of which are given below.

Sensing system configuration

The sensing system (Fig. 2) consists of a sensor controller and linear position sensor, which has a touch probe at its end. The detection sensor incorporates a transducer which converts data obtained by the touch probe to voltage signals. The sensor controller consists of the signal converter, a processor, memory and interface. Their functions are as follows (Fig. 3):

- Signal converter - after eliminating the noise, added to the sensor output signals during transmission, using a low-pass filter, the signal converter amplifies the signals, converting analogue signals to digital signals before sending them to the processor.

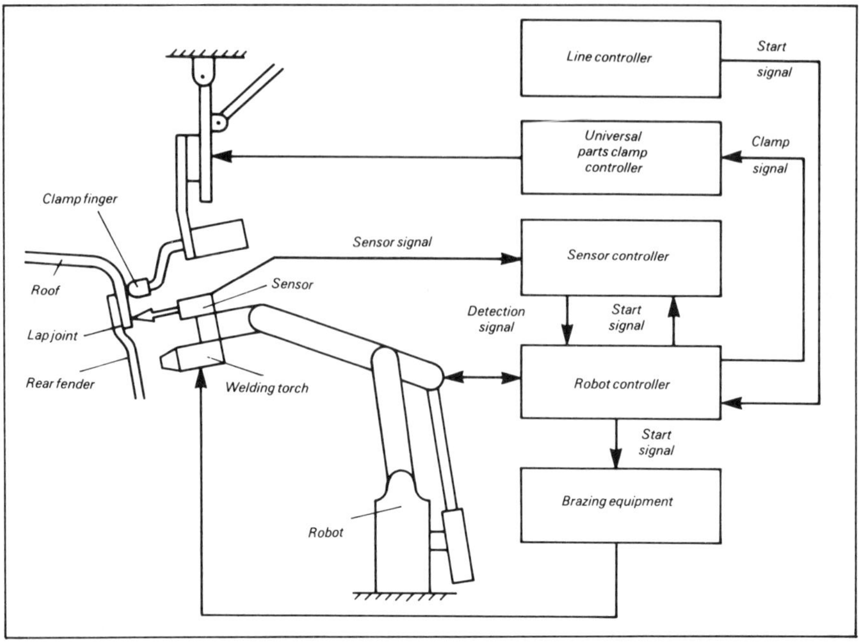

Fig. 2 System configuration

- Memory – the detection parameters are stored in the memory which keeps the established parameters even if the power is turned off.
- Processor – the processor, incorporating an 8-bit CPU, processes the sensor output signals from the signal converter according to the predetermined detection parameters in order to detect a lap joint.
- Interface – the interface serves for transmitting signals between the sensing system and the robot. The interface can thus receive a detection start signal from the robot, or send a lap-joint detection signal to the robot.

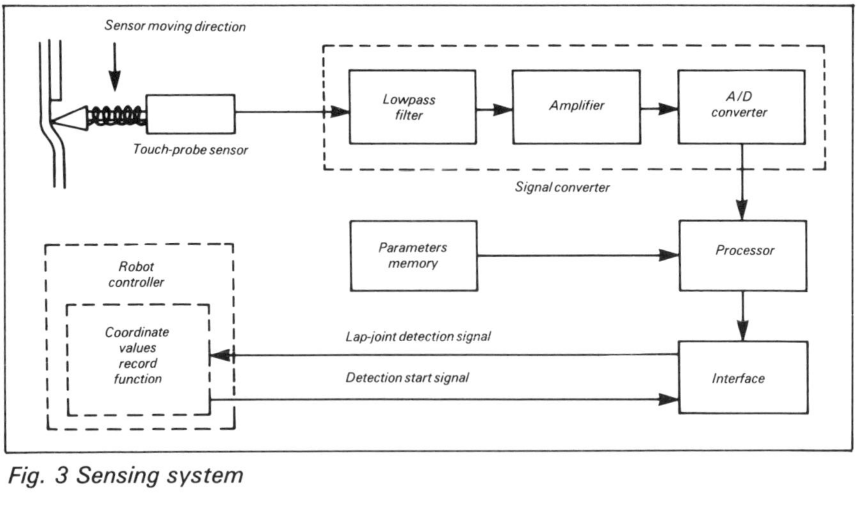

Fig. 3 Sensing system

Detection principle

When the robot moves the sensor over the surface of the car body panels the sensor probe position will vary according to the shape of the surface. The sensor probe outputs a waveform according to the surface of car body panels as shown in Fig. 4. Since the waveform abruptly goes down when the sensor probe reaches a lap joint, a lap joint can be recognised by checking where the different sensor output signal exceeds the predetermined negative threshold level. This causes a protrusion or foreign matter on car body parts to be recognised as a lap joint since the differentiated sensor output signal may exceed the negative threshold level when the sensor probe reaches the protrusion or foreign matter (A, Fig. 4).

However, to prevent mistaken recognition, any excess of the differentiated sensor output signal compared with the negative threshold level, immediately after the differentiated sensor signal has surpassed the positive threshold level, is neglected. Since a differentiated sensor output signal caused by a smoothly waving car body part or robot vibration is smaller than that obtained when the probe reaches a lap joint, an appropriately determined threshold level allows only differentiated sensor output signals obtained by lap joints to be recognised.

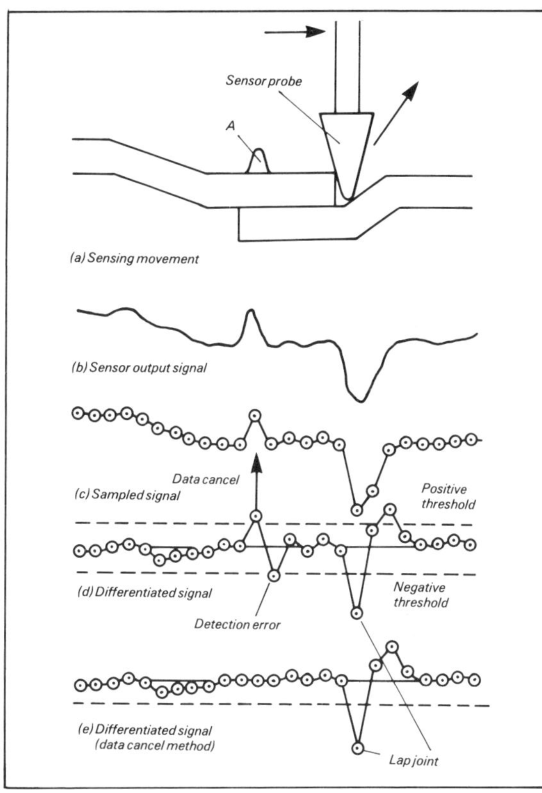

Fig. 4 Detection principle

The sensor output signals are first digitised by the signal converter, then differentiated, i.e. the signals are sampled at a specified interval and a difference in level between two adjacent sampled signals is regarded as a differential value.

If a differentiated sensor output signal exceeds the positive threshold level, which is caused by a protrusion or foreign matter on the car body part, a negative transition of the sensor output waveform will not be differentiated. This prevents a protrusion or foreign matter on a car body part from being recognised as a lap joint.

If the sensing speed is excessively increased, the sensor output signals may not be sampled when the sensor probe reaches a lap joint (Figs. 5 and 6) causing the lap joint not to be detected. However, if the sampling period is shortened, the differentiated values will not be large enough, thus affecting the signal-to-noise ratio of the sensing system. This will confuse the sensing system with regard to whether it should assign priority to the signal-to-noise ratio or to recognition accuracy.

To solve this problem, the differentiated value from the last sensor output value and the value sampled several sampling periods before was calculated, the sampling period being kept sufficiently short. This allows the sensing system to detect a lap joint with a sufficient signal-to-noise ratio without leaving any lap joints unrecognised.

The sensing system detects lap joints at a sensing speed of 50mm/s and detection ratio of 100% at the plant. If the sensing speed is increased to more than 50mm/s, noise caused by the robot's vibration adversely affects the sensing system, impeding practical application of the robot system.

Detection accuracy

When the sensor probe reaches a lap joint, the sensor sends a detection signal to the robot. Receiving the detection signal, the robot recognises and records its current coordinate values. The torch path is calculated

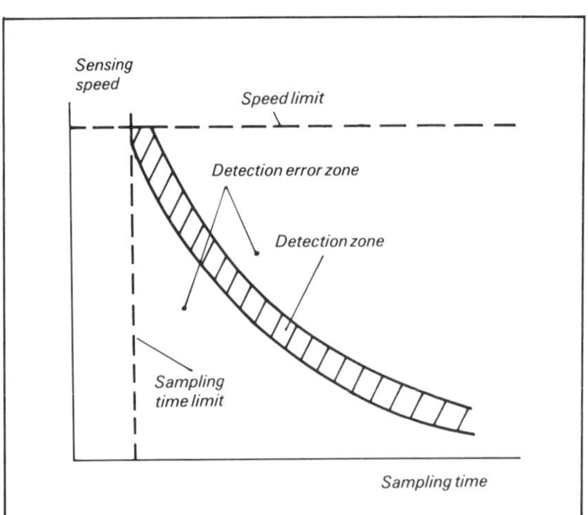

Fig. 5 Sampling time and sensing speed

Fig. 6 Sampling method

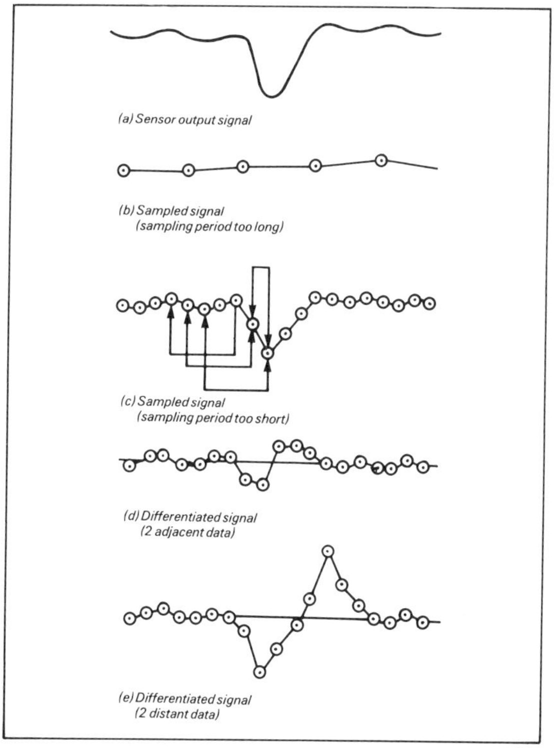

from the obtained coordinate values. Therefore, the error of the robot system is the difference between the sensor position, where the sensor is moved back according to the recorded coordinate values, and the lap-joint position to be detected.

The maximum deviation of the sensor position from the lap-joint position was 0.38mm, based on 50 positioning tests. The deviation is within 0.5mm, as required by the plasma arc welding.

Example of sensing and brazing

The following is an example of the calculation of the brazing path and of brazing operations for the rear section.

Points A and 1 to 5 in Fig. 7 indicate the brazing parts. The robot first detects point 1, then establishes a reference point in the *Y* axis at point 1. Points 2, 3, 4 and 5, which are offset by a certain interval from point 1 and are located on the connecting line of the parts to be brazed, are detected, the coordinate values of each point being recorded.

Point B is defined on the extension of a line segment between points 3 and 4. Point A is also defined on the extension of a line segment between points 2 and 3. The parts are brazed by moving along the straight lines from point A to 3, from 3 to B and from B to 5.

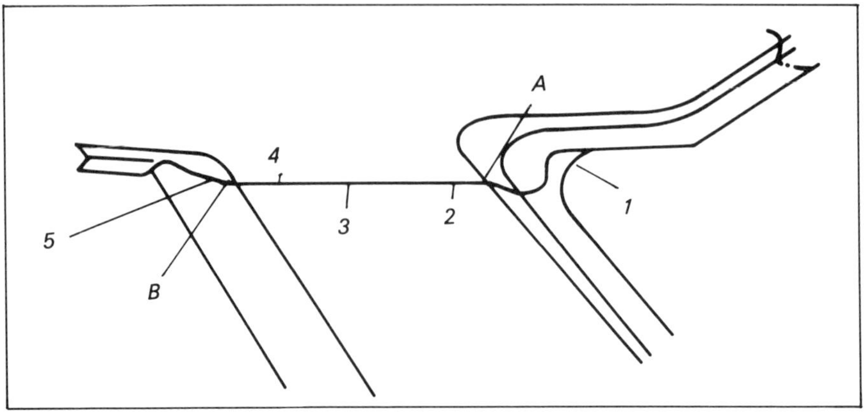

Fig. 7 Detection and welding example

Sensing operation

The deviation of the connection line position from its ideal position consists of *X, Y* and *Z* axis components and components in rotary axes of *X, Y* and *Z* axes. Since the components in the rotary axes are negligible, sensing is determined as follows.

The basic sensing operation is taught to the robot beforehand, so that the robot moves the sensor probe along the surface near the sensing point, maintaining a certain distance between the sensor and parts. The robot moves the sensor to the sensing start position (A, Fig. 8) then pushes the sensor against the part surface until the sensor outputs the predetermined level. The sensor is moved from the sensing start position along the part surface previously specified. When the sensor probe detects a lap joint, the sensor controller sends a detection signal to the robot. On receipt of the detection signal, the robot recognises and records its coordinate values, stops the sensing operation and retracts the sensor.

This method allows the robot to detect and record the actual position of the lap joint without changing the hand direction and distance between the hand and the parts.

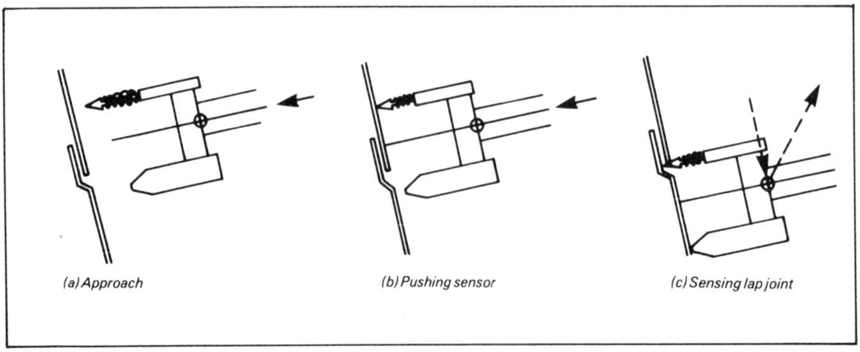

Fig. 8 Sensing operation

Universal parts clamp

Since unsuitable gaps in the overlapping area of car body parts can cause faulty brazing, it is necessary to clamp the overlapped area of the car body parts until sensing and brazing have been completed.

The robot system has a universal parts clamp useful for several models. The universal parts clamp has six fingers, each of which operates independently allowing the fingers to apply uniform pressure on an object despite surface unevenness. The ability of the robot to determine clamping positions on the X-Y table make the parts clamp versatile (see Fig. 9).

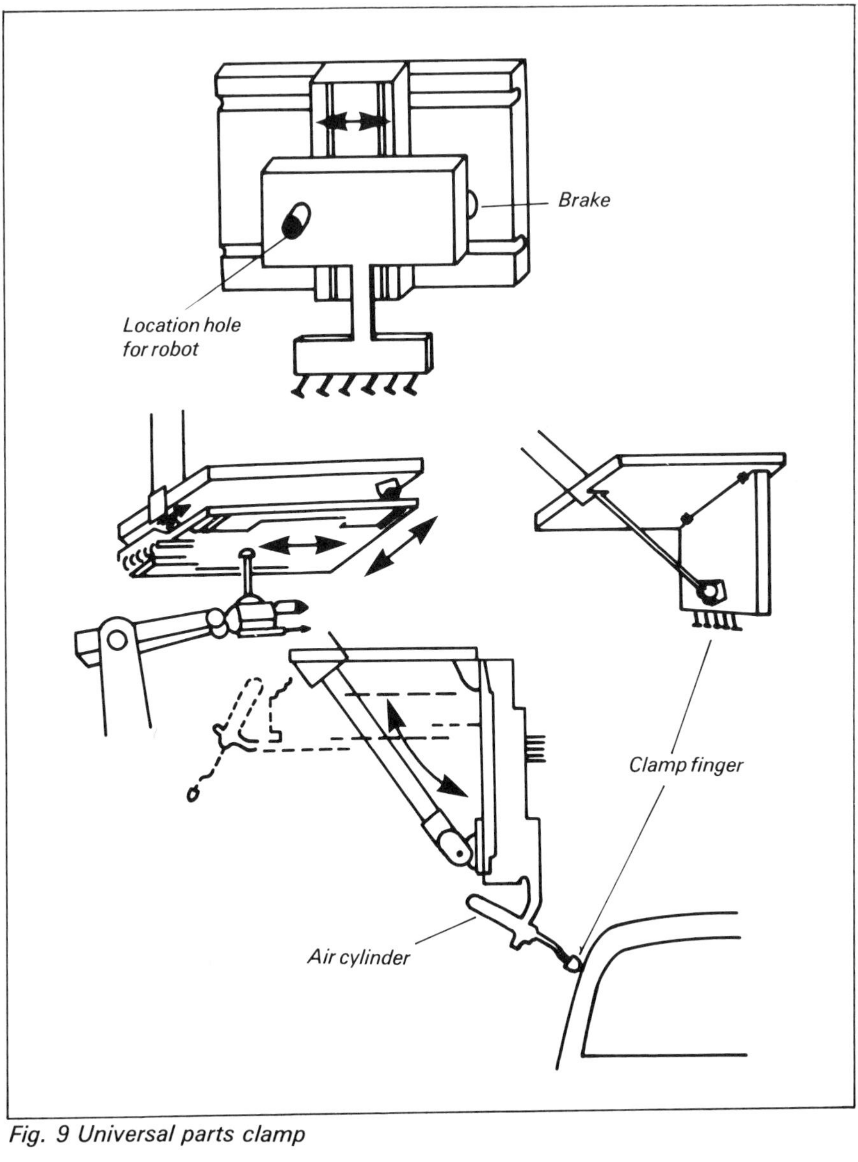

Fig. 9 Universal parts clamp

Concluding remarks

The robot system has a number of advantages:

- Excellent reliability - the lap-joint sensor does not make any misdetection caused by the light source, variations in the shape, colour and surface condition of car body parts.
- High accuracy - the robot coordinates are obtained by simple operations while the sensor probe is kept in contact with the car body part, ensuring high detection accuracy. The maximum deviation of the sensor position from the recorded coordinate values is 0.38mm.
- Low cost - the sensing system consists of a linear transducer, an 8-bit CPU, and a conventional robot with modified software, allowing the robot system to be built at very low cost.

This robot system is presently in operation for two shifts per day. The same robot system will be introduced at the Nissan Motor UK plant in the near future.

Moreover, Nissan has already developed a robot system which operates while following parts on a conveyor. This system is operated in conjunction with the system mentioned above.

References

[1] Loughlin C., Morris J., Rovetta A. and Franchetti I. 1984. Line, edge and contour following with eye-in-hand vision system. In, *Proc. 14th ISIR*, pp. 553-559. IFS (Publications) Ltd, Bedford, UK.

[2] Edling, G. and Porsander, T. 1984. Adaptive control of torch position and welding parameters in robotic arc welding - Examples and practical use. In, *Proc. 14th ISIR*, pp. 359-363. IFS (Publications) Ltd, Bedford, UK.

CONTROL STRATEGY FOR AN ARC-WELDING ROBOT UNDER SENSORY GUIDANCE

W.J.P.A. Verbeek, L.J.H.F. Beckman and G.L. Oomen
Oldelft, The Netherlands

The application of sensors for guiding the path of a welding robot along a workpiece can be implemented in various ways, but most simple methods have some undesirable limitations. A fully satisfactory method, which is worked out in detail, makes use of sensor technology to cope with unavoidable delays in signal processing and robot reaction to external commands. The description is based on the use of the Oldelft 'Seampilot' electro-optical sensor, which measures a surface profile of the workpiece, but the principle of operation is generally applicable.

It is generally agreed that the application of industrial robots for arc welding can be widened substantially, if suitable sensors are available for guiding the weld-torch along the seam. Various sensor approaches have been proposed for this purpose, ranging from tactile guidance to the analysis of the parameters of the welding arc and to advanced electro-optical seam profile measurement and associated intelligent data interpretation.

If a sensor of the latter type is applied, it will look at the seam somewhat ahead of the actual welding position, and at a first glance, this may seem to create undue complications in the utilisation of the sensor data for closed-cycle guidance of the robot.

A more refined analysis reveals, however, that a look-ahead sensor has some remarkable benefits, and that a certain amount of look-ahead is even indispensable in order to compensate for the delayed reaction of present-day robot controllers to external commands for adjustment of the actual tool path. The situation is not unlike that of a driver steering a car, required to look ahead to be able to react in time.

In any practical system consisting of a (welding) robot and a look-ahead-type sensor, the actual look-ahead distance is governed by geometrical factors and unrelated to the controllers delay; it therefore has to be coped with in some generally applicable fashion, if the look-ahead sensor is to provide accurate tracking of the seam by the welding torch. In this paper, several approaches to the problem are discussed, and one fully satisfactory method is described in detail.

Look-ahead sensor geometry

The look-ahead-type sensor which is considered here by way of an example is the Oldelft 'Seampilot' (Fig. 1). This has been described in detail elsewhere [1,2] so that only those features which are relevant to the present discussion are summarised.

The Seampilot is a device for measuring a surface profile of a workpiece in a plane perpendicular to the direction of the robot motion. (Figs. 2 and 4). It obtains that profile by scanning the workpiece with a narrow laser beam and sensing the reflection from the workpiece surface with an array of detectors. As can be seen from Fig. 3, the index number of the detector element which receives a maximum of reflected laser radiation is indicative of the distance towards the workpiece surface.

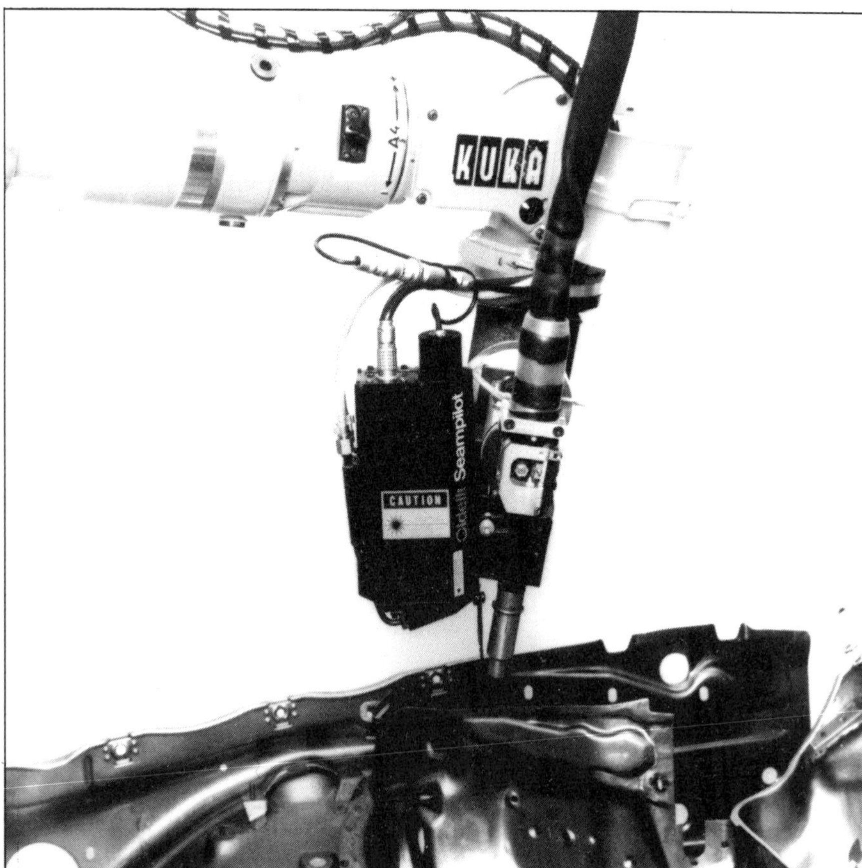

Fig. 1 The Oldelft Seampilot

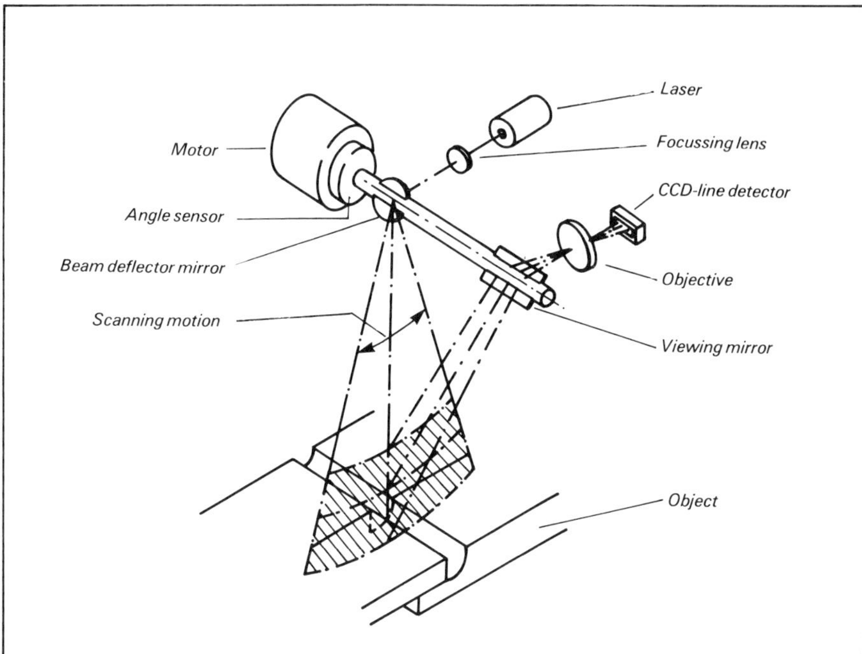

Fig. 2 Principle of operation of the scanning surface profile sensor Seampilot

In actual practice, the instrument measures ten depth profiles per second, each consisting of 200 distance measurements. It covers a fan-shaped area of about 60 × 60mm with a resolution of 0.2–0.3mm. The raw depth profile data are interpreted by an associated computer, which extracts the salient features of the profile, in particular the position of the centroid and the attitude and the cross-sectional area of a weld seam within the field of view, and of a type to be specified by the user.

When mounted on a robot arm, the welding torch and the sensor constitute what is referred to as a tandem vector. Six coordinates are required to fully specify the position and attitude of the tandem, three relating to the tool centre point (torch tip) position, and three to the direction and length of the tandem vector; the latter being a line connecting the tool centre point to the reference point of the sensor, and perpendicular to the plane of sensor measurement (Fig. 4). The torch-sensor distance is fixed and known; it is typically of the order of 20mm.

The actual seam centroid position is measured by the sensor in two coordinates with respect to the reference point in the plane of sensor measurement, the two coordinates being referred to as Y (lateral direction) and Z (vertical direction) (Fig. 4).

Methods for seam sealant tracking with a look-ahead sensor

For the time being, it is assumed that the tandem, consisting of the weld torch and the look-ahead sensor has been placed by the robot in the vicinity of the actual seam and that the tandem vector is, by and large,

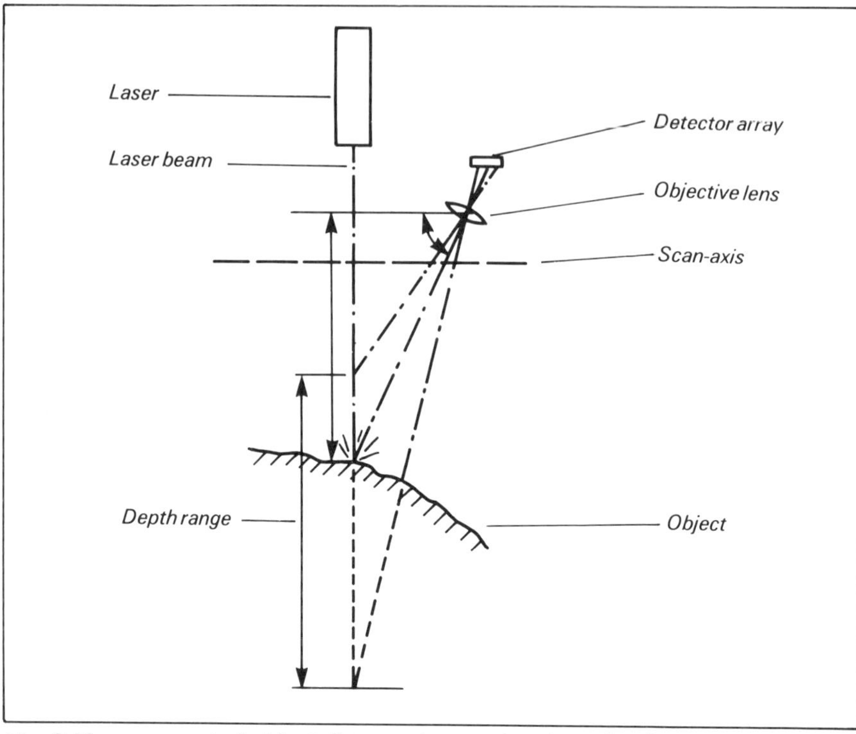

Fig. 3 Measurement of object distance by sensing the reflection of a narrow laser beam with a linear array of detectors

parallel to the direction of the seam. The sensor data *Y* and *Z* (Fig. 4) are the offsets of the actual seam position (in lateral and vertical direction, respectively) in the plane normal to the tandem vector. There are several ways in which these data can be used to provide tracking of the seam.

Direct feedback

The most simple approach is to directly feed the sensor data to the robot to control the robot motions in the two directions perpendicular to the programmed trajectory which is to be the direction of the tandem vector. The method performs satisfactorily provided that the tandem vector is (or can be assumed to be) reasonably parallel to the seam, that is, its application is limited to the correction of positional errors.

Proportional speed

To implement this method, the speed of the robot motions in the directions perpendicular to the programmed trajectory are made proportional to the offsets measured by the sensor, as well as to the trajectory speed. Again the tandem vector must always point in the direction of the programmed trajectory. The method works well for straight seams, even if the actual direction of the seam deviates substantially from the programmed path. Abrupt seam bends are, however, handled rather poorly.

Tractrix

In this method [3], the measured offset data are used alternately to shift the robot trajectory and to rotate the tandem vector, both in such a direction as to minimise the offsets. The method requires a controllable means for rotating the tandem vector about the tool centre point, which is rarely available on existing robots.

Delayed shift

This and the following method use intermediate data storage to bridge the time lag between the passage over any given seam point of the sensor and the torch respectively.

For the delayed shift, the actual seam offset data, measured by the sensor, are corrected, while stored, for any subsequent shifts of the robot arm in the two directions perpendicular to the direction of the tandem vector.

The properly delayed data are then utilised to control further robot motions perpendicular to the programmed trajectory. Due to the correction for intermediate path deviations, the method is quite accurate and suitable to track gross deviations from a 'programmed' path-view of the sensor. While it requires, in its elementary form as described above, that the direction of the tandem vector remains constant in space, an extended scheme could in principle be devised which accounts for rotations as well. A noteworthy advantage of the method is its ease of implementation with existing robot controllers, as it requires only a minimum exchange of data between the sensor and the robot.

Transformation to robot-based coordinates

This is a most general method. Like the previous one, it uses intermediate storage of the sensor data, but prior to storage, these data are transformed into robot-based coordinates. The stored data thereby reflect the actual seam location in space, and thus constitute a list of points on the desired trajectory. This is then fed to the list of points on the desired trajectory which is then used to feed the robot controller for execution in the proper time frame. As the method in effect generates a complete description of the intended trajectory, the programmed path (if any) is completely neglected. It is tempting to claim that the method allows the robot to find its path all by itself, requiring no additional information about the intended trajectory except for an approximate starting point and initial direction.

For the time being, however, it is limited to the generation of positional data only, whereas the tool orientation is taken from preprogrammed inputs. Moreover, it will always be wise to define a (programmed) nominal path and at least to monitor the deviations from it of the actual path.

Key points of this method are the conversion of the sensor data into robot-based coordinates and the insertion of these coordinates into the robot controller for trajectory calculations and control. The remainder of this contribution is largely concerned with these aspects, and with a scheme to perform the necessary tasks through close communication between the signal processor of the sensor and the robot controller. Before discussing this scheme, however, a closer look on the methods of path generation in a typical robot controller is needed.

Generation of trajectory data by a robot controller

The architecture of a robot controller can be generally considered to consist of several functional layers [4,5]. The functions requiring a detailed consideration for the procedures discussed further below are related to the layer which performs the activities of 'path preparation' and 'trajectory interpolation'.

Assuming, that the tool centre point of the robot is presently located at a point a (Fig. 5a), a being a vector of six dimensions, three for describing the tool centre point location and three for describing the tool's orientation. The former will typically be specified as rectangular or Cartesian coordinates, the latter are angular or Eulerian coordinates. Together they give a full account of the tool position and attitude.

In order to move the tool with a specified velocity V, from a to another point b, the robot controller must generate a pattern of intermediate steps at fixed time intervals, specified by the cycle time t. This is done by interpolation in the following manner.

Let A be the distance between two successive (interpolated) points. Then,

$$\mathrm{A} = \text{dist.}\ (x_k, x_{k+1}) = Vt$$

If V remains constant during the trajectory from a to b, then there is a total of N interpolation steps in that trajectory, where:

$$N = \frac{\text{dist.}\ (a, b)}{Vt}$$

N is evaluated when the controller sets out to prepare the trajectory from a to b, a process generally referred to as 'path preparation'.

Now, to calculate the coordinates x_k of the intermediate points, we note that:

$$x_k = a + \frac{k}{N}(b - a) \text{ and}$$

$$x_{k-1} = a + \frac{k-1}{N}(b-a)$$

which can be rewritten as:

$$x_k = x_{k-1} + A$$

with:

$$A = \frac{b-a}{N}$$

By use of this recursive formula, the calculation of the coordinates of successive points, which occurs once per cycle time, becomes rather straightforward, with a minimum demand on computer time. The process described above is graphically depicted in Fig. 5a. The complete procedure includes, of course, a check whether the end point (b) has been reached.

It should be noted that the quantities N, and thus A, are not necessarily constants as they will be used in any somewhat advanced robot controller to manipulate the robot trajectory speed for smooth overall motion. This is, however, irrelevant to the present argument.

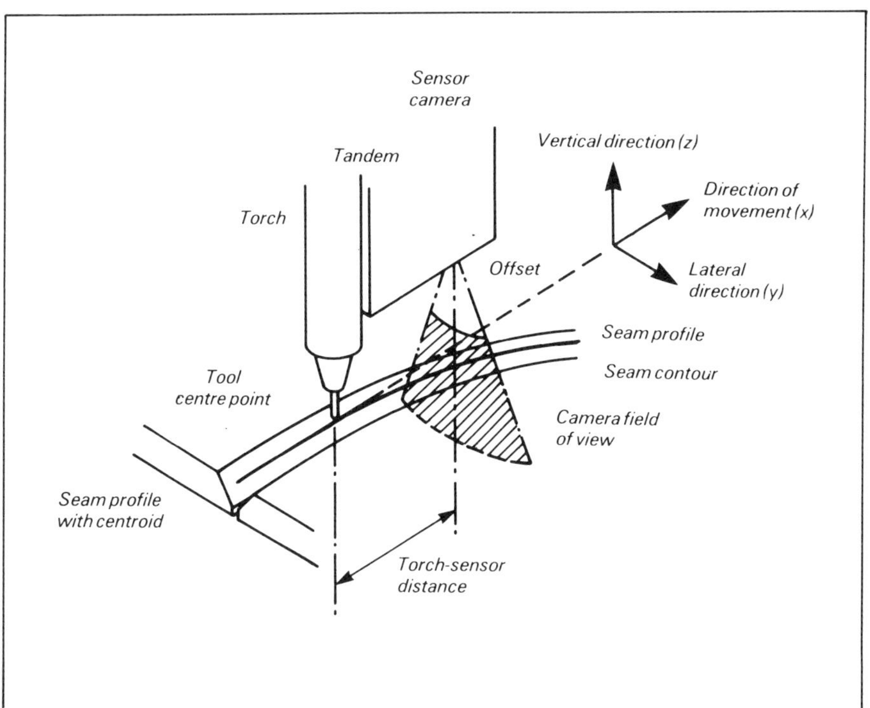

Fig. 4 Geometrical configuration of tool and look-ahead sensor ('tandem')

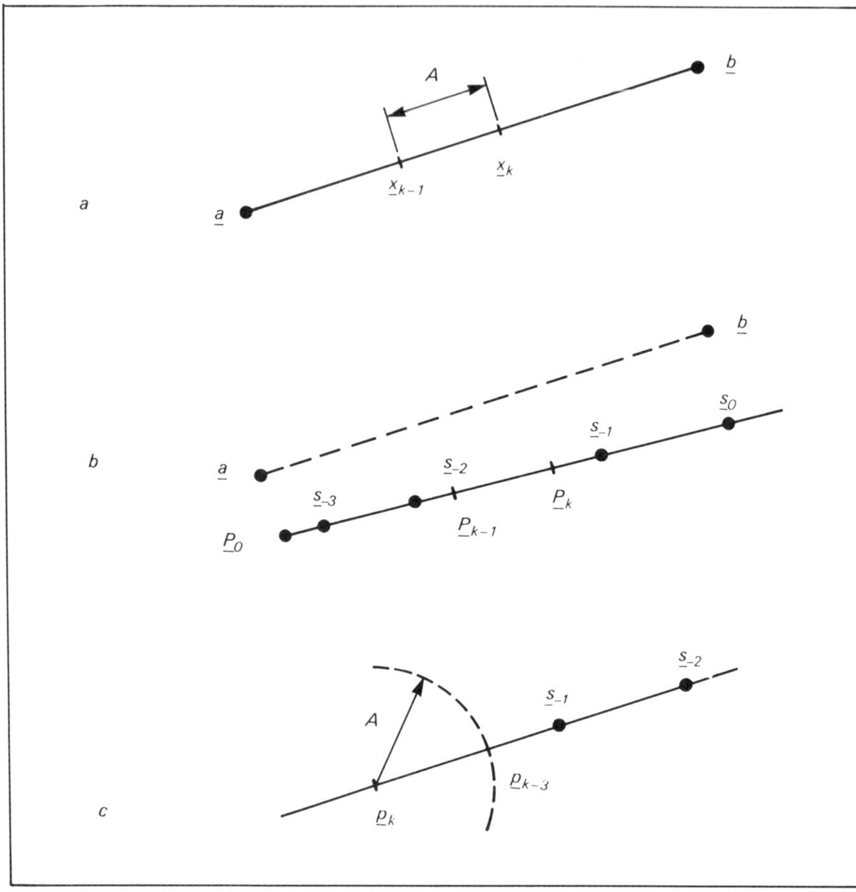

Fig. 5 Principle of trajectory interpolation of a typical robot controller: (a) based on preprogrammed points a *and* b*; (b) for a real object measured by sensor; and (c) for generation of points* p *(see text)*

Real workpiece and relation to a programmed path

The situation pertaining to a real workpiece situation is graphically shown in Fig. 5b. Assuming, that the actual position of the path to be followed along the workpiece has been determined by a series of measurements s_i by means of a sensor. In order to ascertain that the robot follows this path the original trajectory points x_i have to be replaced by a series of new points p_i, and are spaced such as to retain the original spacing of the x_i.

There is no other way to obtain the exact coordinates of the p_i than from the measured points s_i and the measurements clearly must be done at a certain distance ahead of the present position of the tool, making sensor look-ahead mandatory. What is needed in addition, however, is some algorithm which extracts the data for generating a string of properly spaced points p_i out of the measured data s_i which have been collected earlier when the sensor passed along.

This actually involves a dual task, one is to locate a point in proper relation to the present tool position for calculating the next point on the trajectory, and another to cope with the different periodicities at which the s_i have been obtained and at which the p_i are needed.

An appropriate solution is described in detail below. The method requires an intermediate storage of the results of the sensor measurements in a shift register, a device also known as a 'first-in-first-out', or fifo-list. The capacity is such that it bridges the delay between the availability of a sensor measurement and its utilisation by the robot controller. For reasons which will become obvious we require that the sensor data, which typically will be generated in a sensor-based system of coordinates, have been transformed into robot-based coordinates, a process which is straightforward, since the present position as well as the direction and length of the tandem vector are all known.

Method for generating trajectory points

The method described here generates the coordinates of a string of points located on the desired trajectory of the tool along the workpiece with the intention of having the robot controller actually follow that string instead of one based on the preprogrammed path. As stated before, only the Cartesian parts (the three-position coordinates) of the trajectory coordinates, are generated and the appropriate Eulerian parts at their preprogrammed values are kept. Although this is not a necessary restriction, it seems quite appropriate for handling practical situations and simplifies the discussion. It should further be noted that the method has been designed such as to relate to the properties of existing robot controllers rather than requiring essential changes or even replacements on the controller side.

The action really is that of coordinating the activities of the robot controller with those of the (look-ahead) sensor, and the executor of that action (in essence a package of software plus the necessary communication links) may be refered to as the 'coordinator'. The actual physical place in which the coordinator is implemented is immaterial. Depending on individual situations, the preferred location may either be within the robot controller or in the sensor electronics, or even in a separate package.

For a closer look at the method return to Fig. 5b and assume that the robot is presently in location p_{k-1} so that the coordinates of the next point, p_k must now be generated for execution during the next cycle of the robot controller. Clearly p_k needs to be at a distance of A from p_{k-1} but in the direction of the closest measured point s_i beyond that distance A (Fig. 5c).

The first step is therefore to check for the position of the next measured point (s_j) waiting in the shift register, relative to the position of p_{k-1}. If it is closer than at distance A, it is discarded and the next consecutive s_j is called up for which the check is repeated. Once a point at not too small a distance has been located, the proper p_k by an interpolation formula, can be calculated such that

$$p_k = p_{k-1} + A \frac{s-p}{\text{dist}\,(s-p)} = p_{k-1} + p$$

Note that P is vector in the direction of the string of points p (and s) and of length A.

The procedure for generating a new p_k is repeated during each robot controller cycle. It determines the rate at which the shift register is emptied (Fig. 6b).

On the other hand a procedure is needed to fill (or feed) the shift register. This is shown in Fig. 6a, which includes the previously discussed routine and runs once per controller cycle time.

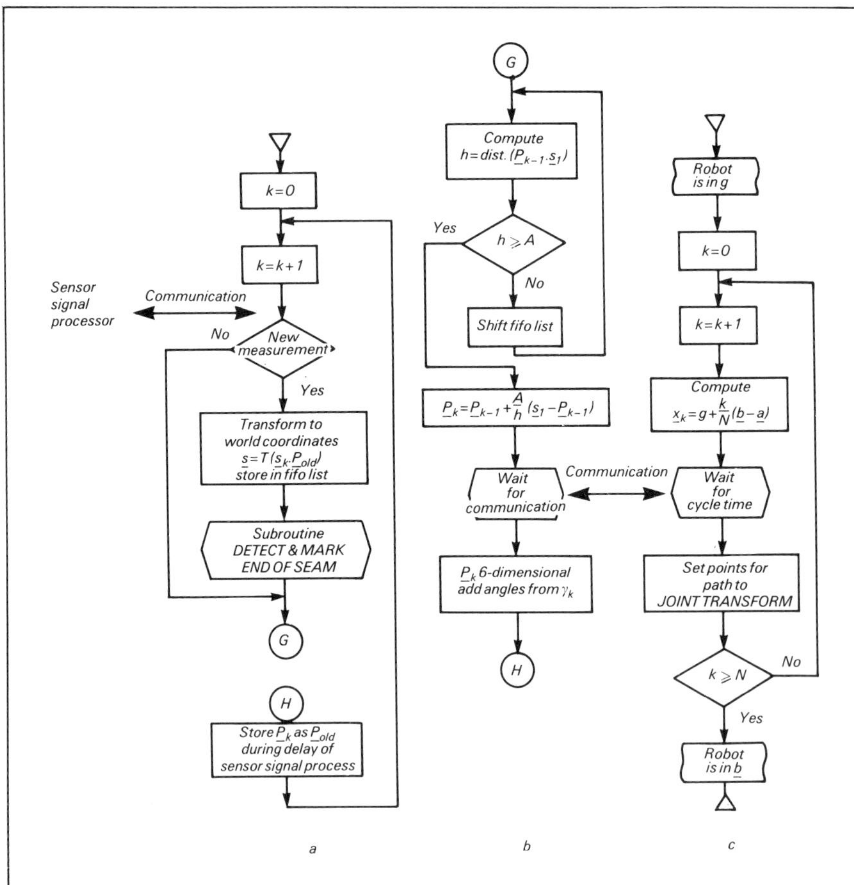

Fig. 6 General flow chart for the activities of the 'coordinator': (a) transformation and storage of measurements; (b) generation of new trajectory points; and (c) interpolation branch of robot controller

The procedure starts with an enquiry - whether new sensor data have been delivered. Obviously, if the sensor operates with a fixed internal measurement cycle, new data may or may not have shown up, depending on the relative cycle rates of the controller and the sensor. No action is taken if there is no new sensor data, but if there is, this data is immediately converted into robot-based coordinates. This implies that a record of the actual tandem position and angular direction at the time of the measurement is kept in store until the measured results become available from the sensor electronics, to ensure that all data are properly time matched. In Fig. 6a, the appropriate tandem position data is referred to as 'p_{old}'.

Having been transformed into robot-based coordinates in this manner, the data are then entered into the shift register. It is interesting at this point to give a closer look at the actual contents of the shift register. As a result of the above procedures, the shift register contains the robot-based coordinates of as many points of the path along the workpiece as are measured over the interval between the present locations of the tool and the sensor, i.e. a three dimensional record of the upcoming trajectory.

Taken as a whole, this set of data allows, in principle, to extract much more relevant information about the workpiece than has so far been - and will be - discussed and utilised in the present context, including information about the slope of the path and its future changes or the end of the seam. These and similar features can give access to a higher level of performance through intelligent processing, indicated in Fig. 6a by a block 'subroutine' but not elaborating in any detail.

Finally, the procedure for transferring the results of the above activities of the 'coordinator' to the robot controller is shown in Fig. 6c. Only those parts of the robot controller loop have been included, that serve to generate point coordinates by trajectory interpolation, and are affected by communication with the coordinator. It should be noted that all standard controller activities are retained and that the essential effect of the communication with the coordinator is just the replacement of the Cartesian part of the programmed trajectory by sensor based data.

Concluding remarks

It has been shown that an advanced sensor can be used to generate data for guiding a robot precisely along a workpiece, the position and orientation of which may necessitate a deviation from the preprogrammed path. It is mandatory that the sensor looks ahead of the present tool position. The processing of the sensor data is such as to keep utilising the procedures implemented in typical robot controllers and thus to retain its known features.

Limited use is made of preprogramming a 'nominal' path, mainly in the area of tool orientation and, if so desired, acceptance limits to deviations. The package of functions which coordinate the actions of the robot controller and the sensor may reside either in the sensor electronics or in the robot controller. The proposed approach is quite general and has the potential for future inclusion of functions of a higher level of intelligence.

References

[1] Oomen, G.L. and Verbeek, W.J.P.A. 1983. A real-time profile sensor for robot arc welding. In, *Proc. 3rd Int. Conf. on Robot Vision and Sensory Controls*, pp. 659-668. IFS (Publications) Ltd, Bedford, UK.

[2] Beckmann, L.J.H.F. 1984. Electrooptischer Sensor fur die Automatisierung des Lichtbogenschweissens. *Technisches Messen*, 51:259-263.

[3] Huber, C. 1984. Sensor-based tracking of large quadrangular weld seam paths. In, *Proc. 14th Int. Symp. on Industrial Robots*, pp. 387-396. IFS (Publications) Ltd, Bedford, UK.

[4] Barbera, A.J., Fitzgerald, M.L. and Albus, J.S. 1982. Concepts for a real-time sensory-interactive control system architecture. In, *Proc. 14th Southeastern Symp. on System Theory*, April '82.

[5] Meier, C. 1983. Sensor technology with the 'Robot Control M'. In, *Proc. 3rd Int. Conf. on Robot Vision and Sensory Controls*, pp. 481-488. IFS (Publications) Ltd, Bedford, UK.

LASER SENSORS FOR ARC WELDING ROBOTS

P.G.Davey, J.W.Barratt and A.R.Vidler
Meta Machines Ltd, UK

The justification and requirements for robot sensors to enable welding in all common metal gauges are discussed, together with an outline of the design of the MetaTorch 200 and 500 systems. Recent developments are outlined that extend the capability of the systems to deal with the full range of joint types met with in MIG/MAG welding practice.

The world market for arc welding robots continued to grow steadily during 1985 even if, as with several other applications, not quite up to the more optimistic forecasts. A June 1985 survey by Prudential-Bache[1] showed US sales increasing by 20% a year to some $60 million in 1985, probably representing about 100 units a month. The February 1986 survey by British Robot Association[2] showed nearly 40 a month being installed in Germany alone. In Japan, the market showed signs of saturation, still increasing, but probably not attaining earlier estimates of about 240 per month. Total sales in Western countries seem to be following this trend, with a delay of one or two years.

By far the greatest number of arc welding robots are used for MIG/MAG (metal inert/active gas) processes, mainly in the automotive industry but to a lesser extent in general fabrication. Some use is also made of TIG (tungsten inert gas) and plasma arc processes, especially for precision work in aerospace and related industries. Arc welding now seems certain to become the first large-scale robot application to benefit from continuous adaptive control by sensors.

In Japan, widespread experience has been gained over the last few years with rather primitive, but useful, sensors such as touch-sensing, through-arc and proximity devices. Overall, sensor guidance is becoming increasingly economic in this particular area, partly because the nature of the process is to require only moderately slow speed at the robot end-effector - up to one or two metres per minute - and this ensures that there is sufficient time to carry out the necessary signal or image processing while still keeping the overall adaptive feedback loop stable.

Another important reason is proof, in a number of production situations[3], that the cost of producing parts and fixturing of higher accuracy can often greatly outweigh the cost of adding a degree of sensing and intelligence to the robot cells handling those parts.

Looking to the future it may well be that the advent of sensors which are both cheap and reliable enough, will be the key to unlocking a large additional market segment, especially for robot MIG welding of jobs which are currently impossible due to dimensional uncertainty. The perception, in Japan particularly, is that this additional market will be unlocked when vision sensing becomes cheap and reliable enough to compete effectively with cruder current techniques.

This paper describes recent improvements to the MetaTorch, a unique structured-light laser/camera sensing system originally based on research at Oxford University[4,5]. The system has been extensively developed by Meta Machines Ltd in collaboration with leading arc welding robot manufacturers. An initial quantity of production-assessment units has been sold for evaluation by users worldwide. While a handful of other companies have also announced early models of vision sensors, none has yet won major multiple orders for routine production use, but the unique design features of MetaTorch seem likely to make it the first to reach this goal. To appreciate these, it is important to review some of the functions that a production sensor-guided robot arc welding system must exhibit, and then present the key features of the MetaTorch solution.

Very consistent arc welding by robot has become increasingly important in the automotive industry in order to achieve structural and safety-critical joints in stronger (such as high strength low-alloy steels), and therefore thinner and lighter sheet materials. Meta Machines therefore initially orientated its laser/camera sensor development toward MIG/MAG welding of thin sheet steel pressings, typically 0.8–3mm thick. This MetaTorch 200 was designed to be wholly integrated, and co-axial, with a special welding torch for best possible access to difficult joints mainly using dip transfer processes up to 200–250A.

The authors describe, for the first time, new developments culminating in an alternative design of sensor, the MetaTorch 500, using similar basic principles and largely common electronics and software. This design offers the production engineer a compact sensor handling most solid-wire and flux-wire and flux-cored MIG/MAG processes in dip spray transfer up to 500A or more, which can be simply fitted to the welding torch of his/her own choice.

Requirements for sensor controlled arc welding robots

Sensor guidance does not consist simply of a system for tracking the centre line of a seam and feeding back position correction to the robot to keep the torch on course. In fact this requirement is only one of at least four functions that should be provided –these are:

- Pre-weld – single point search or multiple point search.
- During weld – tracking of seam centre line along arbitrary 3D curves and continuous measurement of gap in order to provide adaptive control of weld parameters.

Particularly in short three-dimensional seams, it may be nearly as important to locate the correct start point at the right position along the seam as across the seam. This may involve a search, generally in three dimensions, and it is particularly important in order to reduce the time taken, that movement of the robot during this search should be minimised. A single point search is sufficient to enable the welding robot to find the correct centre line (and initial gap) of each seam before the arc is switched on.

In a multiple point search, two or three points are acquired on the seam and/or on features defining the correct start point prior to welding. In the case of a short seam, this may serve to identify the correct point to stop welding, as well as offering the possibility of re-framing, i.e. displacing the taught path in angle as well as position, to cope with location errors of a discrete part whose dimensions are well defined.

Tracking the centre line during welding has attracted by far the most attention over many years and a variety of sensors including through-arc, eddy-current, tactile and visual techniques have been tried. The difficulty is to make any one sensor general enough to handle an acceptably wide range of different joint types, both for thin materials (say 1–2mm) with relatively small radii along the welding trajectory and for thick plate where a variety of mechanised joint preparations must be filled with single or multiple weld passes.

Measurement of, and adaptation to, changes in gap are particularly attractive in situations where it is wished to use arc welding on parts where the seam has not yet been prepared by any machining process, the classic case being automotive pressings. Adaptation to gap is probably best handled by allowing the welding engineer to make entries in a table relating gap, as measured by the sensor during welding, to traverse speed and MIG wirefeed speed, so varying the rate of molten metal deposition.

For thick plate, adaptation of welding conditions to root gap (i.e. spacing at the bottom of a prepared joint) is particularly critical in order to achieve a finished joint with good metallurgical properties. For later passes using weaving, or multiple parallel runs, it is most important to provide adaptation of welding conditions to the remaining cross-sectional area (fill volume) of the joint and this information also must be provided by the sensor.

The MetaTorch approach

From early in the research stage certain major advantages were identified in using an area detector within the camera, rather than alternatives based in a line camera (or other position-sensitive device) and mechanical scanning. The key advantage is to eliminate all moving parts within the sensor attached to the welding torch. It is obvious that no system involving fine mechanical movement offers such a high potential for rugged, reliable operation as a solid-state area camera chip.

MetaTorch designs use stripes generated by two solid-state lasers to provide structured light illumination of the joint being welded. The area camera detects line segments in the field of view produced by scattering of the laser light from the metal surface. Another key advantage in using an area device is greatly enhanced potential for searching for more complex features, which may locate start and stop points for the weld. Searches can be made over a relatively large area with no physical movement of the robot at all, saving time at the pre-weld stage - and larger area searches require only a few movements before the wanted features are found.

A second crucial feature is what has come to be known as 'vision teach'. This consists of a semi-automatic phase following initial (manual) teaching of the weld path trajectory on the nominal component. The object is for the sensor system to have available to it during welding not just the nominal trajectory but the nominal three-dimensional or terrain, at all points along the seam. This gives important advantages in two areas:

- More accurate tracking during welding along sharply curved seams - radii of curvature of, say, 40mm or less.
- More resilient image analysis techniques made possible by correlating vision data received during welding, with prior knowledge of the expected terrain at each point along the seam.

To gain these major advantages, a price must be paid. This is that MetaTorch must have an interface providing continuous information flow in two directions while tracking the weld - from robot to sensor in order for the sensor system to have the best estimate of where the welding torch currently is along the nominal taught path; and of course from sensor to robot providing appropriate gap, lateral, and stand-off corrections at a rate of 10–20Hz.

Significant effort over the last year on the part of Meta Machines Ltd and various robot manufacturers has led to the completion of suitable two-way interfaces on four leading types of arc welding robot. The most recent has been the implementation of searching, finding, re-framing and tracking in the new KAREL language provided in GMFanuc Robotics Inc.'s current controllers. Work is now underway on interfaces to several European process robots.

The third key feature of the MetaTorch approach is the relative ease with which the optical magnification of the camera system can be varied. In this way, increased tracking accuracy down to 0.07mm, for example, can be traded for a smaller total field-of-view, ideal for precision welding of accurately machined parts. Conversely, a field of view up to 40mm can be provided with a minor reduction in accuracy wholly acceptable when welding 20mm plate in, for example, components for earthmoving machinery.

Recent improvements

MetaTorch 200

The design of the MetaTorch 200 has achieved a compact sensor system in the vicinity of the welding torch. The design of the entire laser/camera sensing head, including water cooling, spatter and fume protection within a slender 60mm diameter cylindrical housing (Fig. 1) has now become well known and not only provides great ruggedness but the ability to carry out welding in very difficult restricted areas particularly inside complex pressings and difficult corners.

Recent work has extended the range of distance over which the camera image remains sharp, has enhanced the range of light levels along the line segments in the structured light image which can be handled within a single frame, and has provided continuous interpolation of illumination along the weld trajectory. Most important is a software technique to extend the range of different joint profiles on which accurate centre-line and gap information can be provided.

For the automotive manufacturing engineer, the effect of the camera improvements is such that the sensor can operate efficiently (both in searches, and in tracking) over nearly twice the range of sensor-to-workpiece distance and on a wide variety of metal finishes and profiles around the seam itself. The effect of variations in as-rolled surface finish, die lubricants and handling between press shop and body shop can be largely eliminated. Much sharper local curvature of the sheet metal near the weld line as well as burrs and flash can be tolerated.

In order to deal with a wide range of joint profiles, particularly those arising from curved pressings, seams to be welded by MetaTorch are classified using a new taxonomy incorporated into the vision analysis software. This taxonomy is based on the curvature of the surfaces close to the seam, and on the angle of intersection of the curves locally at the joint.

The surface on each side of the seam can be declared as straight, convex or concave, leading to a corresponding curvature of the projection of the laser stripe as seen by the camera (Fig. 3). During the vision teach process, the actual curvatures and slopes of the tangents to the curves are measured, and the relevant parameters stored. The joint profile can comprise any combination of these features; thus, for

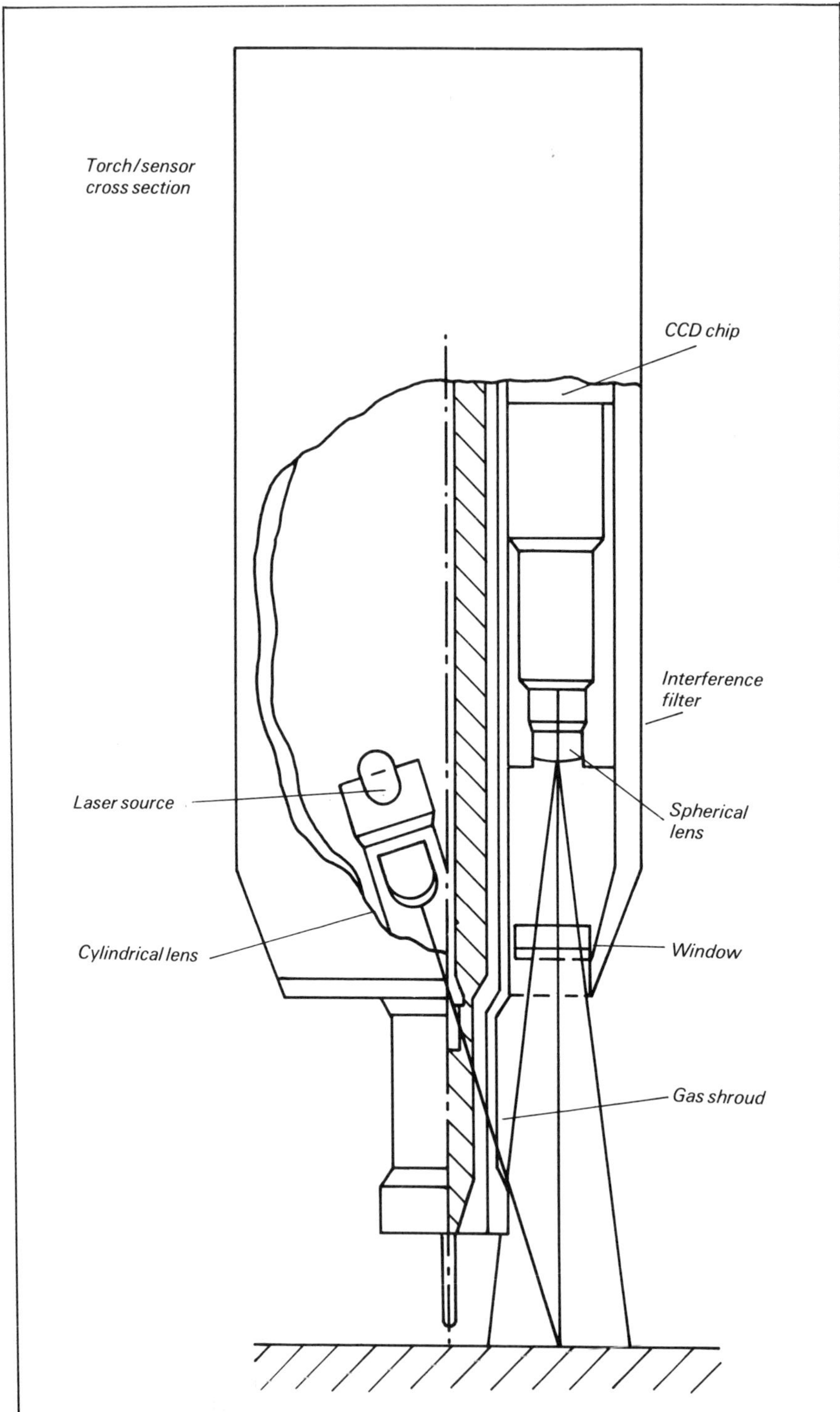

Fig. 1 MetaTorch 200 : simplified cross-section

Fig. 2 MetaTorch 200 welding automotive suspension component

example, the joint type shown in Fig. 3b is defined as concave-straight, while that in Fig. 3c is convex-convex.

The angle of intersection of the tangents to the curves locally at the joint is used to complete the definition of the joint profile. The angle can be used in the region of 90°, as for fillet and coach joints, or about 180° for lap and butt joints, or 270° for external fillet joints. Butt joints are distinguished from lap joints by the fact that the two tangents are not just parallel, but also co-linear. Thus the profile in Fig. 3b is a lap joint, while that of Fig. 3c is a fillet.

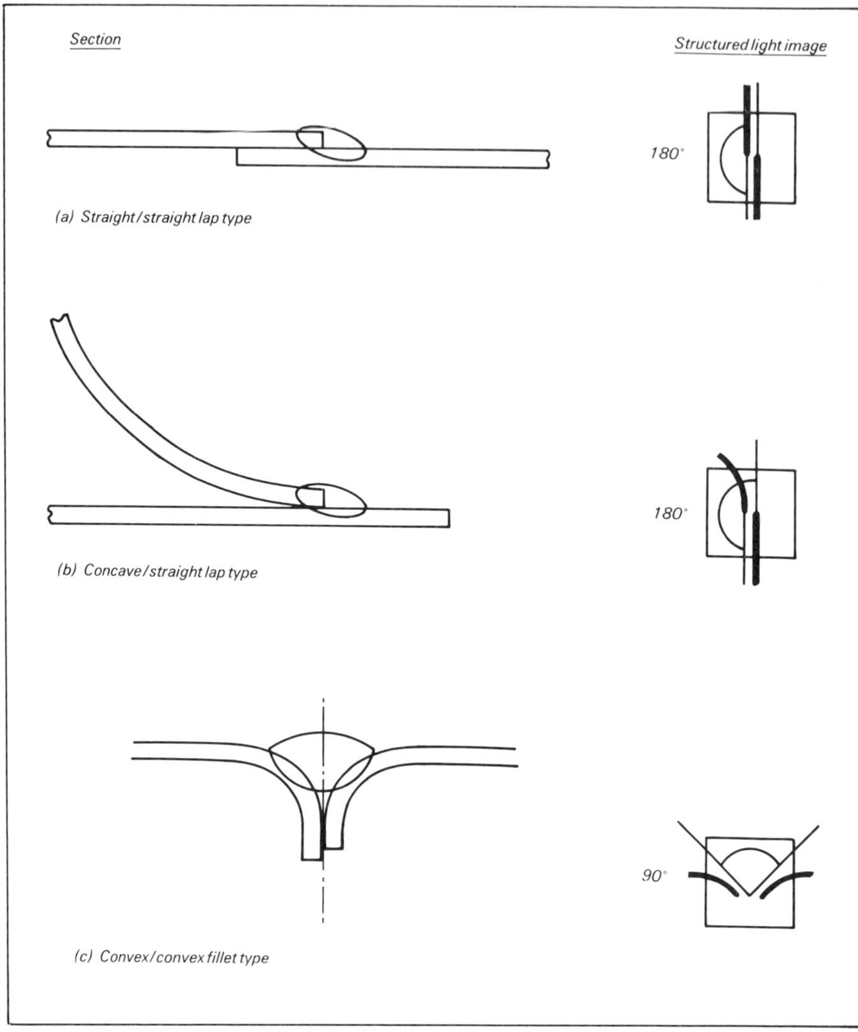

Fig. 3 Classification of joint types in thin pressings

The vision software is now able to analyse joints where one of the surfaces is quite short, so that the laser stripe projection does not extend to the edges of the image.

An optional system for fully automatic maintenance of the laser/camera viewing window has also been developed. This comprises a simple pneumatically operated station mounted anywhere in the robot's working space which provides two functions:

- Indexing of the (annular) glass window to bring a clear area under the aperture.
- Ejection of glass window (in re-usable front cover assembly) and automatic insertion of next from pallet holding stock of six assemblies.

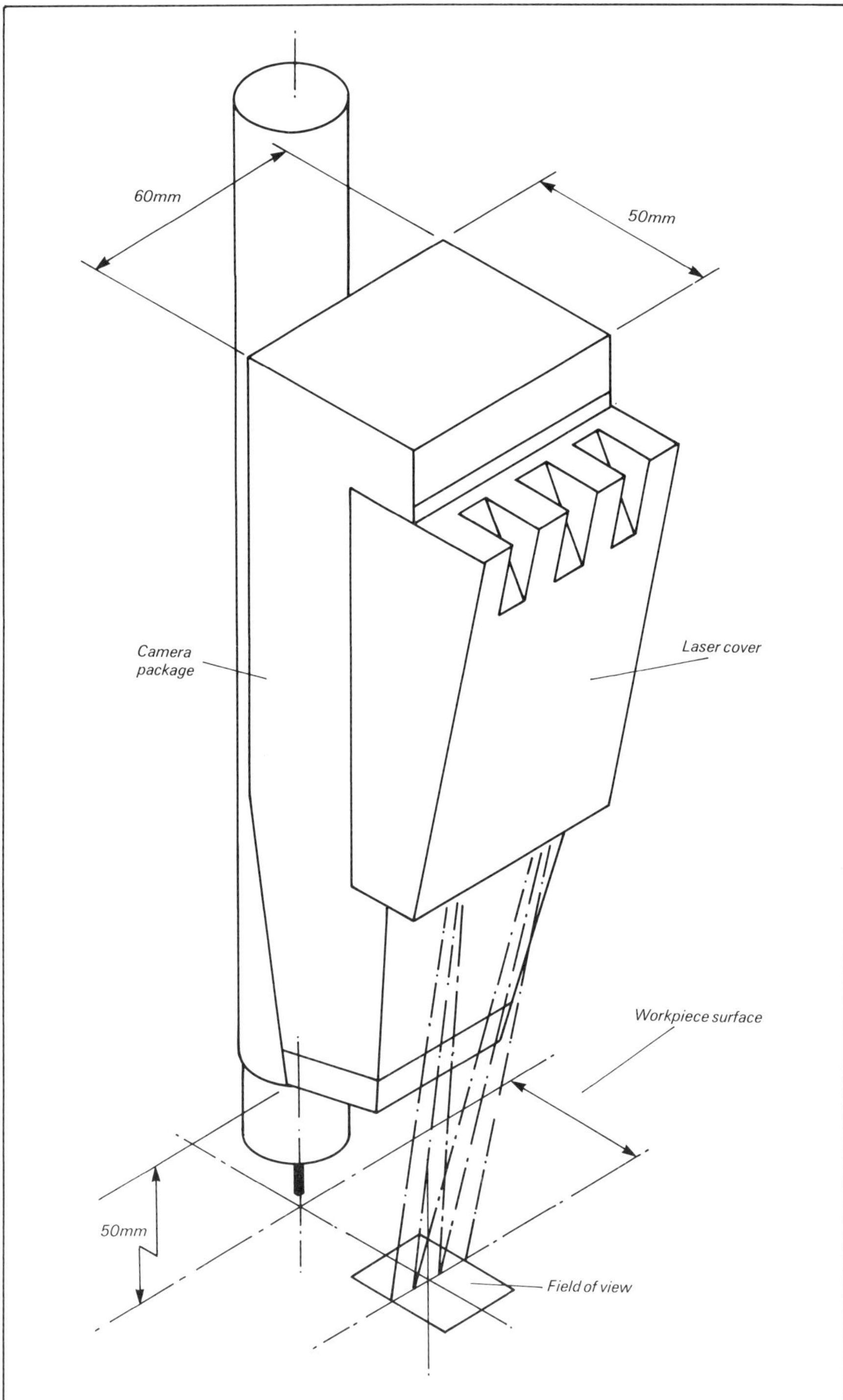

Fig. 4 Outline drawing of MetaTorch 500. Separate sensor unit mounts ahead of standard welding torch

Fig. 5 MetaTorch 500 mounted on GMF S110 arc welding robot

Consistent, high integrity fusion welding is one of the automotive engineer's keys to reducing weight of load-bearing pressed sheet steel assemblies, pending future developments of structural assemblies using plastics and adhesive. The above hardware and software developments together mean that MetaTorch 200 can extend robot arc welding to cover the wider range of seam profiles found wherever complex pressings must be joined.

MetaTorch 500

This is a wholly new sensor head design based on the same structured-light principle. It consists of a separate sensor head which can be fitted, either rigidly or through an optional auxiliary rotator mechanism, to any existing watercooled welding torch whose outer diameter is less than 28mm. It is designed to handle most MIG/MAG processes using solid of flux-cored wire with welding currents to around 500A, 40V.

The MetaTorch 500 package (Fig. 4) occupies a minimum volume ahead of the torch, allowing welding up to internal box corners. It is fully hardened against increased heat, smoke, spatter and glare from the arc. A separate umbilical contains power supply, interlock, control and signal return connections along with separate cooling water and clean air supply. Like the MetaTorch 200, the 500 sensor head contains only solid-state electronic and optical components, with no sensitive fibre-optic or scanning devices. It can be used at once with all robot types for which a MetaTorch 200 interface has been established. Fig. 5 shows a 500 prototype mounted on a GMFanuc S110 welding robot.

The electronics package, image-processing hardware and low-level software are identical for both models. An automatic calibration package takes care of differences due to MetaTorch 500's increased field-of-view, stand-off and lead distance (normally set to 40mm nominal). Higher-level software is offered to handle, for example, V and J preparations on plate to 20mm thickness and calculate fill volumes. Software for multiple-pass work is under development.

MetaTorch 500 is also expected to find applications in heavier automotive work so that existing software to handle the full variety of joint types between pressings is also provided.

Concluding remarks

Sensor guidance can provide a cost-effective solution to problems encountered in robot arc welding if the sensor can operate efficiently over the wide range of metal finishes and joint profiles encountered in practice. Advances in the hardware and software of MetaTorch products, as discussed above, will ensure coverage of a full range of arc welding sensor requirements extending from thin sheet metal pressings through to 10–20mm plate.

References

[1] Conigliaro, L. and Chien, C. 1985. *CIM Newsletter*. Prudential Bache Securities. June 24 1985.

[2] *BRA Robot Statistics*, February 1986. British Robot Association, Bedford, UK.

[3] Barratt, J.W., Clocksin, W.F., Davey, P.G. and Vidler, A.R. 1982. Visually guided robot arc-welding of thin sheet steel pressings. In, *Proc. 12th Int. Symp. on Industrial Robots*, pp. 225-230. IFS (Publications) Ltd, Bedford, UK.

[4] Clocksin, W.F., Davey, P.G., Morgan, C.G. and Vidler, A.R. 1982. Progress in visual feedback for robot arc-welding of thin sheet steel. In, *Proc. 2nd Int. Conf. on Robot Vision and Sensory Controls*, pp. 187-200. IFS (Publications) Ltd, Bedford, UK.

[5] Clocksin, W.F., Bromley, J.S.E., Davey, P.G., Vidler, A.R. and Morgan, C.G. 1985. An implementation of model based visual feedback for robot arc-welding of thin sheet steel. *Int. J. Robotics Research*, 4 (1).

3

SEALING AND SPRAYING

The application of sealing compounds and the spraying of protective coatings is a tedious and inhuman activity which thankfully is now being replaced by industrial robots. The robot can be equipped with intelligent systems able to enhance its performance and give an improved end-product in a process which requires little human intervention, except in the important area of software development – apart from, of course, implementation and commissioning. The use of lasers and vision-guided robots are expected to help advance the technology into other associated areas of application.

SECOND GENERATION ROBOTS FOR ADHESIVE BONDING AND SEALING

E.D. Lawley
Gaydon Technology Ltd, UK
(formerly BL Technology Ltd, UK)

Robots have been used successfully for the deposition of adhesives and sealants in component manufacture. The current applications have been limited by the fluid handling systems which were available. This paper discusses the needs for improved control techniques including the use of programmable flow control, flow rate monitoring and real-time adaptive flow control. These features are now becoming available to the applications engineer, and will greatly increase the role of the robot in bonding and sealing applications throughout industry.

Adhesives and sealants are finding increasing applications in manufacturing industry, primarily due to the much improved performance now offered by these materials which can be achieved on a variety of substrates.

In the aerospace and weapons industry they are used extensively and have enabled engineers to make full use of joining dissimilar materials. This sector of manufacturing industry, although cost-conscious, can devote considerable effort to the assembly sequence and ensure that each assembly is correctly prepared and that the correct amount of adhesive is applied to the specified area.

Timing is important in any manufacturing situation but it is vitally important in the high-volume mass production environments such as the automotive industry. It is essential to keep process times to a minimum without affecting product quality. Any increase in process time inevitably leads to higher costs.

The introduction of adhesives has led to a substantial reduction in the amount of ‘conventional’ joining methods such as welding, riveting and bolting, which have given way to bonding. Within the automotive industry the situation has remained substantially stable over the last five to ten years. Improvements in adhesive and sealant technology have not been matched by increases in either the number of applications or the quantity used.

The methods of applying adhesives and sealants have changed little over the years with the majority of material being applied manually, in both small and large organisations.

Robots

The advent of the industrial robot was viewed with some scorn by the adhesive technologists, partly through ignorance and partly due to the types of robot which were initially available. The early robots were more suitable for the resistance and arc welding applications rather than for applying adhesives. The robot suppliers did little to change this viewpoint and concentrated on the lucrative welding market.

However, in recent years, both users and suppliers have realised the potential for the robot in applying adhesives and sealants. The correct application is essential to the short-and long-term performance of the adhesive. Incorrect application can lead to poor performance and, in extreme cases, failure.

However, to maintain a consistent condition it is essential that the maintenance engineers have a full understanding of the system and can quickly and accurately identify faults. Failure to do this will result in repairs or modifications to one part of the system to compensate for a malfunction in another part.

The robot has the potential ability to carry an applicator consistently around a defined path at a specified speed, eliminating the basic shortfall of a manual system. The robot can achieve this and most of the early robotic adhesive/sealant applications required little extra of the robot. It was quickly realised that its relative inflexibility caused new problems, not least of which was the limited control over material flow afforded by the robot.

The only flow control available to the process engineer or programmer was the limited ability to change robot speeds within the program in order to vary the size of the deposit. This approach gives limited control but is a compromise between what is required, the robot dynamics and available process time. Any change in operating conditions necessitated the re-tuning of the program to achieve the accepted condition.

The results achieved by this technique, whilst superior to that of a manual system, still had shortfalls in terms of quality and consistency; and the manual content in reprogramming to material changes placed unacceptable demands on the already stretched work loads of the engineers.

Slight material surge at the start of the deposition cycle was overcome either by attention to robot speeds or damping of the applicator response. Even with these limitations a generally higher standard of deposition was attained than had hitherto been possible, albeit at the expense of process time.

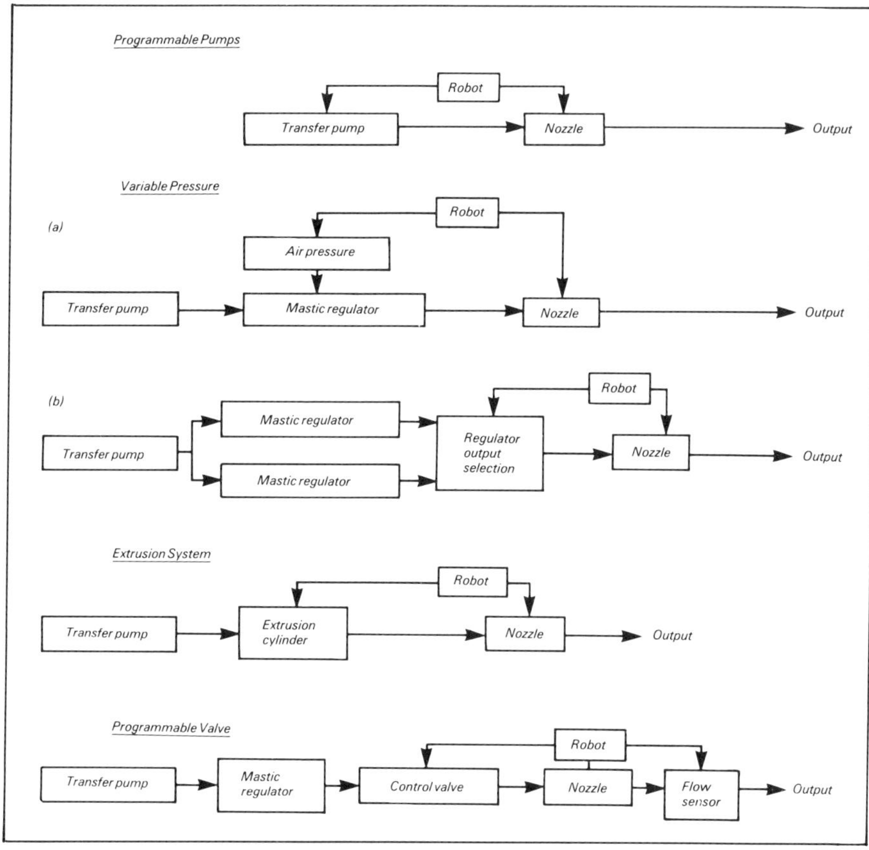

Fig. 1 Schematic comparison of flow control systems

The robots were not without limitations. The application requires the robot to move in a smooth continuous motion along the defined path without significant deviation and with no jerkiness. Any deviation from this condition leads to irregularities in the deposited bead. It has been necessary therefore, for the robot suppliers to improve the overall performance of the robots to meet these needs. While various claims are made by suppliers relating to performance, some users find it necessary to carry out independent tests to assess the robot's true performance characteristics.

Additionally, the robots must be capable of switching signals to activate the applicators at specific locations. Switching at indeterminate points between program location points is not acceptable, and has resulted in changes in programming software.

The introduction of robots to these applications has been matched by any development by the dispensing equipment suppliers, the equipment being essentially automated versions of manual systems.

Robot systems and the increased demands placed by quality and design engineers dictate that besides the control over dispensing provided by the robot itself, additional control similar to that available with arc welding systems was needed if:

- Robots were to advance in bonding and sealing operations.
- The adhesives arc to be used to their full potential.
- The engineer is to design adhesives applications with confidence.
- Material spillages are to be avoided by ensuring that the correct amounts are applied, rather than excess.

There are several basic systems which may be used to control material flow, these are:

- Programmable pumps.
- Variable pressure regulators.
- Extrusion systems.

Each system has potential but it is essential to use the correct robot interfaces to give the full programmability of the dispensing system.

Programmable pumps

The flow rate at the applicator nozzle can be controlled by varying the rate at which the transfer pump propels the adhesive along the delivery pipe, assuming a constant delivery orifice. While this provides some control compared to the constant pressure - constant orifice methods, the response time to fluctuating needs is slow and there is considerable hysteresis in the system and this provides poor overall control over the output rate. It is possible to model the characteristics of such a system and to include this information into the control circuitry. Nevertheless the overall material is still poor; it is also greatly influenced by the compressability of the material being dispensed and the expansion characteristics of the delivery pipework.

Variable pressure regulator

The flow rate of the adhesive through the applicator is dependent upon the fluid pressure in the adhesive line, therefore by varying the pressure in the adhesive line the flow rate can be altered.

Using an air-operated regulator the setting of the regulator can be changed by varying the applied air pressure. An increase in air pressure has an almost immediate effect on the flow rate, but a decrease requires time for the system ahead of the regulator to depressurise before the flow rate stabilises at the lower level.

A similar result is obtained if two or more regulators are used with each set to give specific output pressure and the outputs manifold to give a single supply of the applicator nozzle.

The method gives flow control and can be programmed through the robot control computer. It is useful only where specific flow is required and it is not required to alter the flow along a given path. Even under these conditions care must be taken when changing from a high pressure

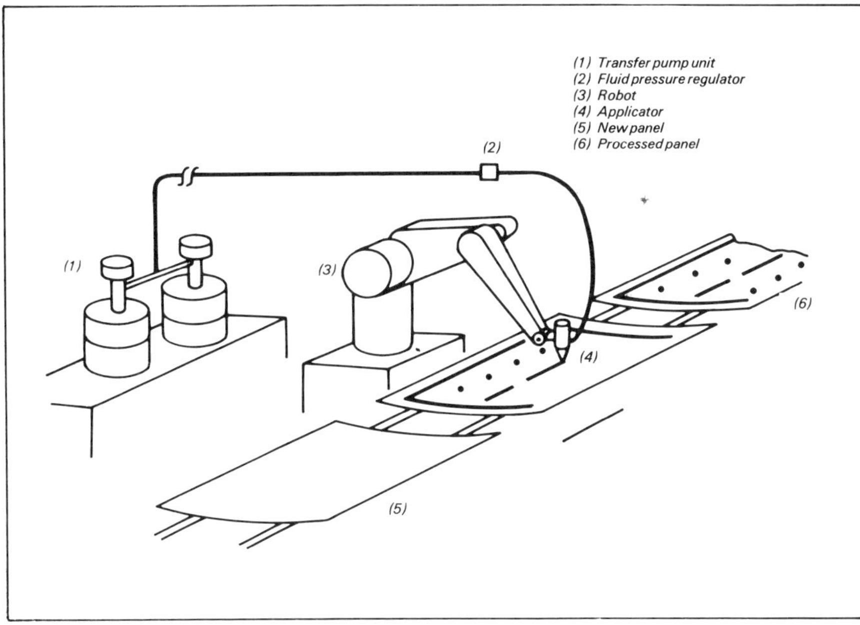

Fig. 2 Typical robot-based application system with automatic panel transfer

path to a low pressure one otherwise problems of surge will be difficult to control.

Extrusion system

Application using extrusion techniques has one major advantage over other methods insofar as it may introduce less shear in the material and therefore may have less effect on the rheology of the applied material. However, some shear is inevitable when the material is transferred from the supply control to the extrusion device.

Two possible approaches exist:

- Extrusion at the point of application.
- Extrusion from a point away from the head, in the feed line.

Extrusion at the point of application. This is achieved using a motorised ram which moves the piston of the extrusion cylinder at a controlled rate. The extrusion cylinder is equipped with a simple on/off valve at the outlet point; the valve is opened during material deposition but closed when the piston is moved backward and the cylinder recharged. Recharging being achieved using a conventional transfer pump system.

The whole extrusion head is mounted on the robot wrist and, being more complex than a simple on/off valve, is heavier which may in turn create robot handling problems.

The size of the extrusion cylinder is influenced by the size of the beads to be applied in a single cycle. In a production application time would have to be allowed to recharge the cylinder between each application cycle. The viscosity of the material also effects overall size of the extrusion head, the stiffer materials requiring a larger drive unit than the more fluid varieties.

A system of this type is commercially available, but currently has a low-power drive mechanism and small capacity. The system is unable to handle the heavier/more viscous materials, as it was designed primarily for dispensing the more fluid silicone liquid gasket type materials.

Extrusion from a point away from the head, in the feed line. The adhesive may be extruded from a point away from the applicator head and feeding the material to an on/off valve at a given rate. A system of this type is now available. It is based on a hydraulic drive unit which drives the extrusion cylinder piston. The size of the cylinder can be readily tailored to suit the specific applications without imposing excessive loads on the robot wrist.

Some delay in response is due to the inherent flexibility of the supply pipework ahead of the extrusion device and the compressability of the adhesive itself. This effect can be reduced by mounting the extrusion cylinder as close as possible to the robot wrist and by using high quality flexible pipework. This gives better control than can be achieved with programmable pumps.

Using this design the size of the extrusion cylinder can readily be tailored to the application needs and it is possible for the drive mechanism to be designed for the material in use without any adverse effects on the robot arm.

Programmable flow valves

The flow of adhesive is controlled by supplying the adhesive to the applicator head at a known constant pressure and then varying the size of the orifice through which it flows. The flow rate then being proportional to the orifice size.

Two basic systems based on the variable orifice principle exist: control at the applicator head and control ahead of the applicator head.

Control at the applicator head

Two systems of this type have been evaluated and both gave excellent results. They work on a needle valve principle, the position of the needle valve being controlled indirectly by electric or pneumatic drive systems mounted directly on the applicator head.

The success of this system relies on the ability to provide the material at a constant pressure to the applicator head, any variation will affect the flow rate. Experience with these has shown that this can be controlled either by the use of a pressure regulator in the supply line to the control valve or the use of a single-acting pump based device designed and proportioned to have sufficient capacity in a single stroke to accommodate the maximum bead volume.

Extensive trials with two variable orifice systems have been carried out. With the first system the valve was controlled through the output lines of the robot controller to a stepper drive mechanism mounted on top of the valve body. Using a binary coded signal the valve could be controlled to give very fine control of the valve.

A similarly designed system using a pneumatic positioner to control the valve opening gives similar results.

The commercially available package provides three to eight preset levels of flow, these are selected through the output lines of the robot control. By modification to the control logic of the valve and using analogue outputs from the robot control, this valve can be changed from providing preset levels of flow to continuously variable control and is therefore comparable to the electrically driven systems.

Control ahead of the applicator head

To reduce the load on the robot wrist and allow the use of larger control valves the variable orifice device can be mounted on the robot arm ahead of the applicator/on-off valve. Whilst this introduces some delay in response, the use of special supply lines between the valve and applicator reduces this to a very low level.

This approach has the advantage over the more compact integral units above, in that it can be constructed from more robust equipment without incurring weight penalties, and due to the larger proportions of the control valve it is possible to manufacture the control needles to well-defined profiles to give any required control characteristic.

A control valve based on this system has been developed by BL Technology (now Gaydon Technology Ltd). This has been developed and designed to be able to handle a wide range of materials and has available a range of 'standard' control needles to give differing operating characteristics.

The system can be controlled through either digital or analogue signals from the robot controller to give either preset levels of flow or continuously variable control. Additionally, to assist programming the equipment uses a simple display device to indicate the flow levels selected.

The equipment is manufactured, by Graco (UK) Ltd, under licence from Gaydon Technology Ltd.

Concluding remarks

Each of the systems discussed requires a mechanism to transfer the adhesive from the supply containers to the control system, and will require to be pumped using conventional transfer pumps. These will inevitably introduce some shear into the material and consequently alter the rheological characteristics of the material when deposited compared with those of the supplied material.

Each of the systems discussed above could be used to control the flow of adhesive but only the extrusion and variable orifice devices have the widest range of applications.

Commercially available and suitable versions of the variable orifice device do exist; they are of both the wrist and remotely mounted versions.

Currently only the one small extrusion system is known to exist, and this has limited power in the drive mechanism and is unlikely to be able to handle viscous materials.

Each application will have its own particular requirements and the user must select the appropriate system to give stepped or continuously variable control over the output rate of the adhesive.

Programming of the total system must, until the next generation of flow control devices are developed, be a slow manual process, tuning the application cycle to meet the needs of the application.

Using real-time flow sensing and an adaptive control loop it will be possible to make the whole system self-monitoring and controlling. With typical robot speeds of up to 250mm/s processing of the flow information must be fast to enable real-time correction to be made.

In some industrial applications, a total cycle time for applying an adhesive bead of 6s is not unusual, with single beads being deposited in less than 0.5s.

The advent of the programmable flow control system enhances the robot's capability and will assure the immediate future of the adhesive bonding and sealing robot.

STEREO VISION FOR ROBOTIC SEAM SEALANT SPRAYING

P. Johnstone
Automatix International (UK) Ltd, UK

This paper outlines the theories and technical approaches to vision-guided robotics and artificial vision relevant to seam sealant spraying. While both structured light and stereo vision techniques are considered, stereo vision is considered more appropriate and is therefore described in greater detail. Finally, an actual stereo vision based robotic seam sealant spraying system is described.

To prevent water ingress between the steel sections of fabricated car bodies, it has always been necessary to provide a seal between various panels and subassemblies. Traditionally, this seal has been achieved by manually brushing the sealant into the appropriate seams and joints. This tedious process often results in an inconsistent seal and poor utilisation of seam sealant material. However, recent advances in seam sealant dispensing technology, robot control and artificial vision have been combined to provide an alternative approach.

A more acceptable method of applying the sealant is to force a free-flowing sealant material into the car body joints from a narrow high-powered jet. The inherent penetrating characteristics of the narrow jet promote a much improved quality seal than that provided by manual methods. Also, as the sealant may be applied exactly where it is required, i.e. between adjoining body panels, this method is more efficient in terms of material usage.

This approach to sealant dispensing requires accurate positioning of the spray gun (to within ±1mm) combined with consistent, free-flowing movements. An articulated 6-axis continuous path robot is an ideal prime mover for this application as it offers the correct dynamic characteristics in conjunction with a potentially automated solution.

However, the economics of automotive manufacture dictate that positioning a car body to within ±1mm on an indexing conveyor is prohibitively expensive. In reality, car bodies may be positioned in a station to within typically ±75mm by conventional overhead indexing conveyors. Hence, to fully automate the seam sealant process, a vision-guided robotic approach is required.

The theory of vision-guided robotics

Homogeneous transformations

When considering the theory behind vision-guided robotics, the usual way to describe a position and orientation in three-dimensional space, and the basis for all the mathematics described, is with homogeneous transformations. These are 4×4 matrices:

$$\begin{matrix} nx & ox & ax & x \\ ny & oy & ay & y \\ nz & oz & az & z \\ 0 & 0 & 0 & 1 \end{matrix}$$

The right-hand column gives the position of an object or reference frame relative to a base reference frame. The 3×3 matrix in the upper left-hand corner describes the orientation of this object relative to the base reference frame. Each column describes how an individual axis in the object's reference frame appears in the base reference frame. For example, the left-most column [*nz ny nz*] describes how the *x*-axis in the object's reference appears in the base reference frame.

The 3×3 orientation matrix is an orthogonal matrix. This means that it transposes its inverses implying that the columns (and rows) are all unit vectors and mutually orthogonal. It is easy to show that the product of homogeneous transformations is also a homogeneous transformation.

An example should make these clear. Consider the homogeneous transformation below:

$$\begin{matrix} 0 & 0.707 & -0.707 & 1 \\ 0 & 0.707 & 0.707 & 2 \\ 1 & 0 & 0 & 3 \\ 0 & 0 & 0 & 1 \end{matrix}$$

This describes a reference frame or an object located at [1,2,3] from the base reference frame. It indicates that the *x*-axis in this new reference frame (or in the one defined by the object) points along the base *z*-axis. It also says that the *y*-axis is a 45° unit vector in the first quadrant of the *x-y* plane of the base reference frame and that the *z*-axis is a 45° unit vector in the second quadrant of the *x-y* plane. This reference frame, like all the ones to be used, is right-handed.

Using a base

Consider a path in a part. This is a series of positions and orientations for the end-effector of the robot. It can be described as a series of homogeneous transformations, one for each taught point. Let us denote the path relative to the robot by *P(R)*. The path can also be described in a reference frame relative to the part, the world origin for us. In this reference frame, let us denote the path by *P(W)*. If, in turn, we know the position and orientation of the part relative to the robot, which we will call *B* (base), then we can relate the two expressions for the path as follows:

$$P(R) = B{:}P(W)$$

where ':' denotes matrix multiplication done for each point in the path.

This equation allows us to redefine the path relative to the robot by changing the value of one variable *B*, the position of the world reference frame relative to the robot. Assuming that the path relative to the world reference frame is the same for all parts (at least within the tolerance of the process), then we can teach the path once.

For future parts, all that has to be done is locate the world reference frame relative to the robot. Then, the same path can be played back, defined in the new world frame. In this case *P(W)* will remain the same, but *B* may have changed; therefore the robot's motions may change to accommodate the new definition of *B*. In some applications the world reference frame may be defined as the part itself; examples include a stator winding to be inserted into a generator housing or a non-standard component to be inserted into a printed circuit board.

Artificial vision

The idea of vision-guided robotics is to use artificial vision to gather the information which enables the robot to determine where the part is. In practice, the vision system determines where points in the part are. The robot system then uses that information to build a reference frame, the base. The first time through, the user teaches the path relative to that base, producing *P(W)*. Then, for each subsequent part, the vision system and robot jointly locate the part, producing a new base. The robot then plays back *P(W)* relative to this new base, moving differently from how it was taught as the part's position may dictate.

To enable the robot to build a base, the vision system must provide three-dimensional information to the robot. The number of points needed depends on the assumptions about how the part will move. If the part is expected only to experience translation, then only one point is needed. If the part is expected to lie anywhere in a plane, then two points are needed. Finally, if the part is expected to experience arbitrary rotation and translation, as experienced by car bodies on an indexing conveyor, then three points are needed.

Two techniques are used to generate this three-dimensional information. They are stereo and structured light. In both cases, the cameras must be calibrated before they can be used using calibration blocks. This means that measurements must be taken which enable the vision system to take two-dimensional image space coordinates and determine the three-dimensional positions to which they correspond. The use of calibration blocks is described later.

Both stereo and structured light produce three-dimensional locations in the reference frame of the cameras. In order to make this information useful, where the cameras (or, specifically, the calibration blocks) are relative to the robot needs to be known.

Fixed or mounted cameras

Three configurations are possible. The simplest is fixed cameras. Here, the cameras are bolted to the floor. To determine where the calibration block is, three known locations on the block are touched with a pointer attached to the end of the robot. This enables the builing of a reference frame describing the position and orientation of the calibration block relative to the robot. In this case, if the part is located at $X(C)$ in the camera reference frame and the camera reference frame is located at $X(R,C)$ the base, or location of the part relative to the robot, will be:

$$\text{Base} = X(R,C) : X(C)$$

The other extreme is tool-mounted cameras. Here, the cameras are located on the end-effector of the robot so it is necessary to build a reference frame describing where the calibration block is relative to the tool of the robot. In this case, if the location of the camera relative to the tool is $X(T,C)$, the location of the tool relative to the end of the wrist is $X(Wr,T)$, and the location of the wrist relative to the robot's internal reference frame is $X(R,Wr)$ (produced by the arm solution), then the base will be:

$$\text{Base} = X(R,Wr) : X(Wr,T) : X(T,C) : X(C)$$

The last case is midway between these two extremes. Here, the cameras are mounted in an intermediate axis of the robot. The equations change accordingly.

Vision-guided robotics, as described here, can be used when the robot's path remains the same in the reference frame of the part; only the part moves. Vision can then be used to determine the reference frame of the part, i.e. where the part is relative to the robot. Stereo and structured light are two ways to do this, but which method is appropriate?

Stereo vs structured light

Structured light

Stereo works by triangulation of an object in space from two camera positions. Structured light uses a single camera and a sheet of light. The light sheet geometry and location in space is known. Only the intersection of world points with the light sheet are considered in structured light. This reduces the world points to a two-dimensional surface which is imaged onto a two-dimensional focal plane. The world points map uniquely into focal plane coordinates so that a function applied to the focal plane coordinates can be used to recover world coordinates.

The key point to note is that there is no correspondence problem when structured light is used. Thus, features are not required to measure spatial positions using structured light. The position of points in space on a featureless sheet-metal surface may be determined using structured light techniques. On the other hand, since the location of a structured light line is normally fixed, there is no guarantee it will go through a given feature. Structured light is most appropriate for gauging parts, determining the shape of a surface and locating gaps and edges.

Stereo

Stereo should be used only when there are clear features which can be located repeatedly. They should preferably be large in order to be located accurately. Gauge holes in thin metal stampings, round spots printed on a box and slots in a surface are typical features used in stereo applications. It is desirable, although not a requirement, that the features used be symmetrical. Feature centroid is commonly used to increase feature location accuracy by taking advantage of improved location accuracy from multiple scan lines passing through the feature. A feature can be assymetrical and centroid location will still work well if the feature always has the same orientation. Table 1 summarises the advantages and disadvantages of stereo and structured light.

Hence, in the case of locating a car body for seam sealant spraying, stereo vision is appropriate, as gauge holes in the car body are good targets. The seams themselves, which may be used if structured light is selected, do not enable the vision system to locate the car as quickly or effectively.

Stereo vision

Stereo vision uses two cameras to generate three-dimensional information just as our two eyes present a three-dimensional view of the world. Similarly, each camera takes a slightly different picture of a scene. Fig. 1 illustrates the general stereo approach. It consists of two parts. The second part is entirely general and mathematical in nature, whereas the first part is application dependent.

Pinhole optics

The mathematical part of the algorithm essentially uses trigonometry to locate the feature in space which corresponds to the two binocular images. The stereo vision algorithms assume a ‘pinhole’ model of the

Table 1 Stereo vs structured light

	Advantages	Disadvantages
Stereo	No special light sources. Can determine location of object features	Requires two cameras
Structured light	Requires only one camera. Works on featureless surfaces	Needs special projector light sources

camera optics. This means all rays reaching the camera focal plane have travelled through a common point referred to as the optics pinhole. Since any light ray striking the focal plane has gone through this single, fixed point in space (the pinhole), a focal plane location together with the pinhole location determine a unique line in space. A point which is imaged by a pair of cameras will determine a pair of lines in space. These lines must intersect in space at the original object point. Physically this can be thought of as triangulation of the object's position, as illustrated in Fig. 2. The stereo code takes as input certain camera calibration coefficients which essentially characterise the cameras in space, together with the focal plane coordinates and uses them to locate the corresponding space point.

In practical situations, the lines in space from a common point will not appear to intersect due to noise and deviations from pinhole optics in a real camera. Thus, point location is done in a least squares sense; that is, the point will be chosen which minimises the sum of the squares of the normal distances to each of the triangulation lines.

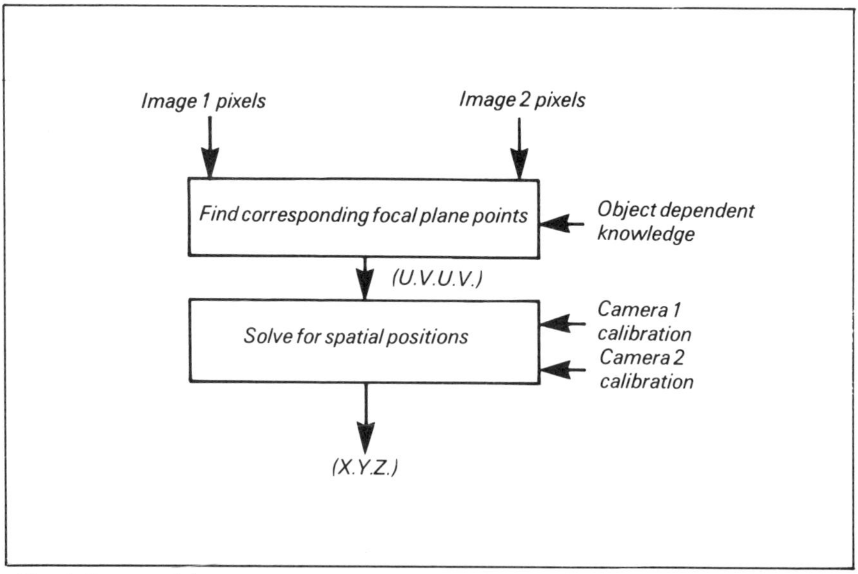

Fig. 1 Stereo vision processing consists of two parts

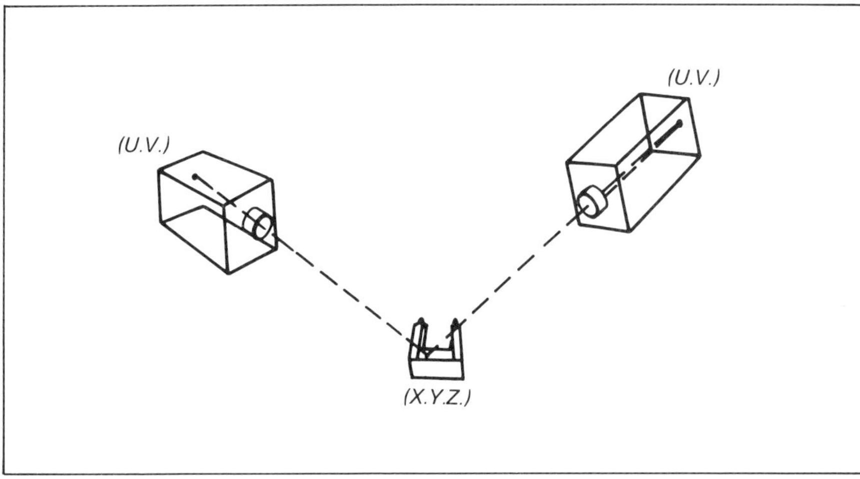

Fig. 2 Stereo projection of a point into focal planes

Calibration

To perform stereo reconstruction, it is necessary to know the characteristics of the camera in space – the position and orientation of the effective pinhole – as well as other camera dependent parameters such as the focal length of the lens and the size of a pixel in the focal plane.

One characteristic of the camera calibration procedure is that it is not necessary to attempt to measure any of these parameters directly. Instead, the cameras take pictures of an object with distinctive features on it in known positions, called a 'calibration block' (Fig. 3). When the apparent positions of these features in the image plan are compared with their known positions on the calibration target in space, all of the

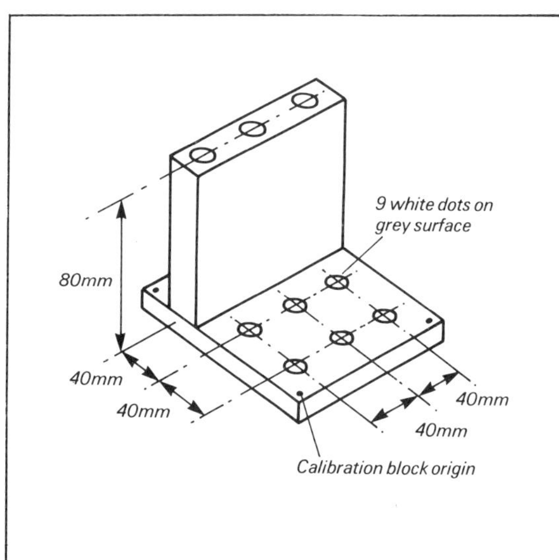

Fig. 3 Stereo calibration block

necessary information about camera calibration can be inferred without any explicit measurement. This technique greatly simplifies the calibration process.

The camera images then have to be processed to correctly extract the relevant features. A pair of digitised images of the same object from different camera positions are processed to find features which are to be located in space.

As described earlier, the features used in locating the car body are gauge holes. Each of the three gauge holes is found in three-dimensional space by a stereo camera pair. These three points are sufficient to determine the position of the rigid car body in space for seam sealant application (Fig. 4).

Having described the various technologies and theories that provide an automated solution to the process of seam sealant dispensing, an actual case study is now examined.

Seam sealant spraying system

The system described is for applying seam sealant to car body wheel wells. The system is situated on an overhead indexing conveyor which positions car bodies to within ±75mm.

The primary components of the system are as follows:

- Four AID 900 articulated robot arms with six degrees of freedom, electric dc servo drive, 15kg payload, ±0.2mm repeatability, 2200mm reach.
- Four AI32 robot controllers with multiple M68000 processors, software based, three RS232 ports and 16 opto-isolated I/O lines.

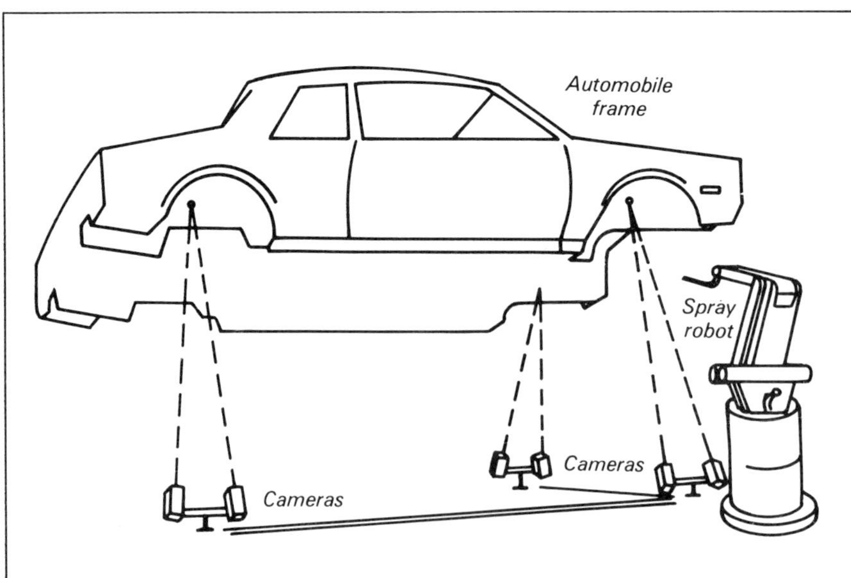

Fig. 4 Stereo location of a car body to direct a robot in seam sealant spraying

- One AV4 vision system with mutiple M68000 processors, software based, 64 grey levels, three 404×256×8 picture buffers, supports eight cameras, three RS232 ports and 16 opto-isolated I/O lines.
- Eight 380×252 pixel solid-state cameras with 75mm lens.
- Four 300W wide-angle floodlights.
- One sealant dispensing system including:
- Four 30:1 Bulldog pumps mounted on 55gal pneumatic rams including air regulator, pumps and ram controls.
- Two double-ram changeovers.
- Four master fluid regulators and high pressure filters.
- Four automatic air assisted airless spray gun with filter spray tips.
- Four accumulators for surge soothing.
- Four flowmeters.
- One programmable logic controller with 4K RAM, 48 input/output lines.

System operation

The eight cameras in four stereo pairs are used to locate four gauge holes in the car underbody frame. The cameras are fixed outside the working envelopes of the robots as the wheel wells offer a very restricted workspace for robot mounted cameras. The hole locations are transmitted to each robot controller through RS232 communication lines through a BTA 524 multiport controller. Each robot controller uses the gauge hole locations to determine the position of the car relative to the robot's learned base. The robots then execute a program to spray sealant along the seams in the wheel well.

The vision system is the master controller over all four robots. Opto-communication between the vision system, the robots and the manufacturing line is done through a programmable logic controller.

The sequence is as follows:

- Car coming into station trips a limit switch which activates the pre-stabiliser.
- Pre-stabiliser clamps car.
- If in automatic mode, cycle start signal is received by the vision system from the line. If in manual mode, cycle start is simulated from the PLC by pushing the 'test cycle' button.
- Vision system activates the stabiliser.
- Vision system waits for 'car stabilised' signal from the PLC.
- Vision system checks to see if robots are ready.
- Vision system sends cycle start to robots and robots begin to move to finished position near wheel well.
- Body style identification is received on the AV4 from proximity sensors and transmitted to each robot controller.
- Vision system takes pictures (while robot is moving) and calculates 3D position of each gauge hole.
- Hole locations sent to each robot.

- Each robot calculates its own base and sprays the wheel well for that car.
- Each robot signals that it is completed.
- Robot cycle complete signals are daisy chained (in the PLC) to signal the vision system to release car.
- Vision system releases stabiliser and checks air cylinder for complete retraction.
- Cycle complete sent to line.
- Total cycle time is 23s.

The primary feature of this vision-guided robotic approach to seam sealant dispensing is very rapid sealant application in conjunction with consistent delivery, reducing the amount of wasted sealant due to excessive application.

Concluding remarks

The robotic seam sealant spraying system described is just one example of the effective use of vision-guided robotics and stereo vision. Vision-guided robotics allows robot work-cells to adapt to changes in the work environment without human intervention. Using artificial vision techniques, such as stereo vision, workcells can be constructed which minimise the need for precise fixturing. The techniques have a solid grounding in theory and also work effectively in practice.

Acknowledgements

Many people have contributed to the success of the project described and to the preparation of this paper, including D. Lees, G. Phipps, W. Smith and P. Trepagnier.

ADVANCED PAINT SHOP AUTOMATION

A.D. Wiley
Gaydon Technology Ltd, UK
(formerly BL Technology Ltd, UK)

The application of robotics and advanced manufacturing technology to surface finishing is a challenging, but rewarding mission. The benefits are well understood, in terms of achieving competitive advantage and improving both the neighbourhood and manufacturing environment. The problems are less well understood, but relate to the complexity of the finishing process. The solutions will be achieved by considering surface finishing as a total system, consisting of process, quality and application methods, automated by the application of sensors, microelectronics and robots, networked into an integrated manufacturing system.

The paint shop presents an ideal environment for automation, as it provides nearly all the criteria conventionally associated with the case for applying automation or advanced manufacturing technology.

These criteria will be examined in more detail but can be summarised at this stage as:

- Quality.
- Productivity and output.
- Environmental considerations.
- Material utilisation.
- Energy considerations.

The paint system can be perceived as two main activities:

- Application methods.
- Processing.

Both of these activities can be studied separately as cases for automation, but in the final analysis, must be interfaced together to provide a total interactive system.

This paper follows a similar line of reasoning, discussing automated process control and management, the automation of application methods, and the interfacing of the process with the application method.

Common to all three cases are the criteria, or motives, briefly mentioned above, and consequently these require some explanation as a background to the paper.

Quality

This is, arguably, the prime factor supporting automation, and affects both the consumer, and the manufacturer.

The consumer expects a perfect, defect-free, cosmetic finish, in a range of alternative finishes, providing the durability to protect investment.

The manufacturer needs to achieve this consistently, on a range of products in a cost-effective manner. This is particularly important in today's ever-increasing competitive environment.

The solution is to minimise any manual activity, and implement flexible automation.

Productivity and output

Both these objectives can be satisfied by the implementation of advanced manufacturing technology or AMT.

Paint shops tend to be manually intensive, particularly in areas such as sealing, preparation, and spraying. The savings in manpower are obvious, with less obvious indirect savings in protective clothing, and so on.

From a business management perspective, paint shops are traditional bottlenecks in the total manufacturing process, often resulting in loss of production output, excess work-in-progress, and unplanned re-work. This results in loss of revenue, excess manning, and excess material utilisation. Almost always, these output problems are caused by human shortcomings. It is clear that automation has a substantial role to play, and the resulting economies must go a long way toward achieving planned returns on capital investment.

Environmental considerations

Paint shops create adverse and hostile environments at two levels: the working environment and the neighbourhood environment. Both of these can be significantly improved by the application of AMT.

The painting process consists of an integration of a range of processes: cleaning, wiping, blowing, sanding, spraying, and so on. All of these can be claimed as being at least unpleasant, and in some instances (such as spraying), hazardous. The simple answer is to automate.

A word of caution, however, is in order: automation is not always the correct solution. On occasion, alternative methods provide a better solution, or even deletion of the process in question. Two examples can be given: use 'non-sand' substrates in place of sanding, and use dipping processes in place of spraying.

Paint shops also can cause adverse effects within the neighbourhood, e.g. paint overspray escaping from exhaust stacks, smells from solvents and ovens, and so on. Automation has a positive role to play in constraining this problem. It exerts more effective controls of material utilisation, thus controlling overspray, and permits the use of environmentally more acceptable materials which might cause difficulty if applied manually.

Material utilisation

A significant cost in the painting operation is materials. An automated system will control material utilisation more closely than manual methods, with consequent savings, which are easily quantified. Often overlooked are the savings that can be made in the use of water by tighter, automated controls.

Energy savings

A paint shop is a significant user of energy in spray booths, heating, chemicals, ovens, and so on. Automation makes significant energy savings, when applied to both process control and application methods. Experience has shown energy saving to be a significant element in the cost-saving equation.

In summarising the justification for applying AMT in the paint shop, it is to be observed that the paint shop probably represents one of the least difficult areas to justify economically, and yet manufacturing industry appears to be slow to meet the challenge of automation in this area. It is probable that this represents a major growth area of the future, which is just beginning to emerge.

Having studied the underlying reasoning, this paper examines some specific examples of where automation can be, and has been, applied within the paint shop.

Automated process control and management

Product quality is of prime importance and this is particularly so in the paint shop. Probably the singlemost important factor contributing to good paint shop quality is process control. This is normally a manual task whereby an operator either checks an instrument or gauge, or takes a measurement. This is unsatisfactory because it is unreliable, manually based and not suitable to automated plant and methods.

The answer is automated process control, which will provide:

- Increased first time OK rates.
- Increased output.
- Total system quality assurance.

- An aid to trouble-shooting.
- Improved warranty figures.
- Cost savings, materials, energy, water, manning.
- Condition monitoring of plant.
- Complements automated manufacture.

Gaydon Technology (formerly BL Technology) has developed and implemented several automated process control and monitoring systems within the British Leyland paint shops. Due to their automated nature, these systems can operate on several models from alarm, specific parameter readings, full system scan, summary of 'out-of-spec' conditions, end-of-shift displays, to graphical displays in colour graphics.

These systems have been installed on pre-tested plants, electro-dip facilities, ovens and spray booths, and may be networked into a total paint shop control system. These can be expanded to include both quality data systems, and body tracking or mapping systems.

A system normally consists of central management terminals and plant-based programmable controllers communicating with each other in a data highway.

A typical central management facility comprises:

- A low cost 16-bit microcomputer.
- Floppy disk drive unit.
- Colour graphics visual display unit.
- Colour printer.
- Black-and-white printer.
- Keyboard.

Programmable controllers are used as stand-alone intelligent devices providing plant monitoring and control functions at shopfloor level.

The programmable controllers communicate their actions to the central management terminals through the data highway link. The system is designed to communicate with a mainframe computer providing information on system status and monitored parameters as necessary.

The plant sensors and control signals are interfaced to the programmable controllers through both analogue and digital input and output modules.

The sensors have been developed to monitor a range of parameters, for example, temperature, pressure, conductivity, flow, voltage, current, levels, turbidity, and so on.

The cost of such a system can range from £150,000 to £300,000 for a total system controlling an automotive paint plant to £25,000 to £30,000 for a relatively simple pre-treatment controller.

Automation of application methods

Almost without exception, the methods used in a paint shop can be described as hostile, and in some instances like spraying required specialised, air-conditioned booths. These manual methods also have a direct effect on product quality. The case for automation is particularly persuasive.

Paint spraying

Paint spraying ranges from the simple application of a single colour to a simple object, to the complex application of many colours to a complex object; for example, motor car bodies. The total system will need to interface component ident systems, auto colour changers, the robot, the conveyor, the possibility of manipulators to facilitate paint access, material process control, spray booth design and operation.

Some basic issues need to be understood and resolved when implementing robotic paint spray. The robot should be seen to complement existing methods, such as reciprocators and/or electrostatic spray.

Tracking a moving conveyor introduces another level of complexity. Consideration should be given to indexing paint cells, or moving the robot with the product. Process control, particularly of paint fluid flow-rates, is important and will vary colour by colour.

Programming is difficult. It is time-consuming, interferes with production, and the current manual systems mirror manual weaknesses in the program. The ultimate answer lies in off-line programming, based upon CAD/CAM principles. Robot selection is of key importance, and the ultimate selection should be based upon a list of criteria and benchmarks derived by the user.

Product design is also a key factor, in terms of access, clean lines, and commonality of materials. Serious consideration should be given to a modular approach to improve access, which is probably the major constraint.

Van interior spraying

The application of paint to van interiors is a good example of a process that gains the maximum benefit from automation, and as such will be examined in some detail. The manual application is an extremely unpleasant task with the operator working in a confined space, with poor extraction and lighting. Gaydon Technology developed and demonstrated a suitable system within its robot development area.

The requirements to be met by the system include:

- Automatic colour changing - 15 colours.
- Automatic model identification - 35 models.
- Ablility to deal with all models including high-roof/long-wheel base derivitives as well as short-wheel base low-roof versions.

Various methods of automation were considered and evaluated, and included:

- Dedicated automation.

- Electrostatic mini bell boom mounted.
- Robots.

Neither of the first two provided the flexibility required for the application. The production system developed was based on:

- Twin robot booths to provide a breakdown contingency.
- Each robot is mounted on the end of a boom.
- Vans are alternatively indexed into each booth and spraying is carried out both as the van is conveyed in and out of the booth to provide the maximum process time.

The vans then proceed into a common booth for exterior spraying.

The benefits obtained included: paint saving, quality improvements, consistent quality, and the removal of the operators from an environmentally unacceptable task.

Undersealing

This is another good robot application. The main problems to be overcome are nozzle blockages and tipwear, body movement, position and tracking, material flow control, programming and maintenance. As in spraying, the optimum solution to automatic underseal is to use a robot system in conjunction with a conventional reciprocator.

Seam sealing

This is another ideal robot target, highly manually intensive, quality critical for leaks and an unpleasant task working inside a moving shell. The main challenges are tracking an accurately presented body seam, and access. The solutions will include simple tracking devices, indexing cells, and improved access via modular construction techniques.

Body cleaning

Here operations are carried out manually either by wiping, usually with a solvent or tacky rag, or blowing with compressed air. They are both critical operations to ensure cleanliness and freedom from contamination. These will be difficult challenges for automation owing to the random nature of the task.

Interfacing the process with the application method

Manual operators, particularly when paint spraying, contribute more than simply applying paint. They are aware of the cosmetic importance of the tasks and are continuously, perhaps subconsciously, assessing the effects of their work, and adapting accordingly. Perhaps spraying more quickly, or slowly, changing the distance, or angle of the gun from the workpiece, feathering the gun, all to achieve the optimum finish.

Total automation will remove this manual, intuitive reaction to change. This will consequently make total automation at most unlikely, and at best difficult.

One solution is to mathematically model the relevant process parameters and their relationship with process defects. This model would then be installed as part of the total automated paint system. It is also probable that a level of process modelling will be required as an element of off-line programming of the painting process.

To achieve this model, the paint process will need to be broken down into subsets, for example, automisation, travel, deposition and flow. In parallel, algorithms will need to be established relating process variables with resultant effects. The total will be combined to produce an overall model, which will provide feedback control for the complete system.

Concluding remarks

This paper shows the application of automation in finishing to be worthwhile, with achievable paybacks, but it is a demanding challenge. A full system approach is essential, supported by the full range of disciplines, the robot engineer, the production engineer, the product engineer, the systems engineer and manufacturing management.

ROBOTS IN THE UNMANNED SPRAYBOOTH

H.F. Pitcher
Haden Drysys International, UK

Simple robots were used to apply finish coats to automobile bodies in 1951. Three-axis robots developed from these early machines coat over 70% of all external surfaces in most modern body paint shops. Initially six-axis robots were used as man simulators but the requirements for painting are not the same as for assembly and welding. Some progress is being made with second generation multi-axis robots to paint the difficult-to-get-at interior surfaces in attempts to remove all production labour from inside the finish spraybooth.

Quite apart from the obvious advantages of eliminating manual operations in unsocial conditions there are worthwhile reasons for attempting to achieve the fully automated unmanned spraybooth, such as:

- Obtaining improved and consistent quality of the finished product on visible and hidden areas.
- Improving first time OK rate (reduction in run, sags, dirt).
- Matching the booth operation conditions to the process requirements without any compromise to meet operator comfort conditions.
- Increasing the choice of coating materials, e.g. toxic risks will not preclude the use of certain coatings.

Automobile body painting

It is proposed to use the painting of automobile bodies as a good illustration of the attempts to achieve full automatic painting in a particular industry, the current state of the art, and immediate prospects, but first it may help to highlight some of the major problems to be overcome.

First, it is current practice in the automotive industry to paint the assembled passenger car body complete with doors, bonnets and boot lids and to apply a cosmetic finish to the so-called 'shuts' and also interior areas such as engine bay, baggage compartments, etc., albeit often with a less onerous standard of finish. This practice was adopted in the late 1930s in the UK and in the 1950s in the USA, mainly to avoid colour matching problems and obtain a better utilisation of plant. With one notable exception, this practice has continued to date.

Secondly, in seeking to optimise inventory the auto manufacturers have tended to reflect customer/dealer style and colour preferences on the production painting schedules with minimal regard to colour batching. It is not unusual to see finished bodies at the unload end of the paint shop conveyor in a variety of colours and body style variations.

Thirdly, with the object of reducing the manual labour required for obtaining the high gloss finish demanded by car owners, the paint manufacturers have developed finishing processes known as gloss-off-the-gun, reflow, and more recently, clear-over-base, which 'lock in' and magnify any 'dirt' in the paint film. This 'dirt' may be present in the plant due to inherent design faults or poor housekeeping, but more often it is brought into the spraybooth on the body, or by the spray operators, or in the paint as pumped from the mix house. Blemishes in the cured paint film caused by settlement of 'dry' overspray under, or on top of, the wet film are sometimes blamed on 'dirt'.

When designing a fully automated system the latter is of particular importance as the order in which the multitude of operations are carried out within the spraybooth is critical and is not necessarily the order which would best suit the automatic painting equipment design.

Finally, it sometimes appears as if the paint shop is the Cinderella of the body assembly plant. Often the easy option is to call for a coat of paint rather than make a small modification to the design of interior trimming materials.

Historical development

Corrosion protection and priming

Fully automatic systems for metal pre-treatment and priming (Fig. 1) were successfully developed and in production in the UK in the late 1940s and early 1950s but it was not until the mid 1960s, with the introduction of electrophoretic painting, that equivalent or better standards of preparation of the body for finish coat painting without the employment of production-line process operatives were adopted universally.

Today, in addition to automatic electrophoretic dip priming there are installations where the inside of structural 'box' members of finished bodies are automatically coated and/or filled with 'waxes' to reduce the risk of corrosion arising from road salt deposits accumulating in crevices and partial enclosures in the underbody. Such wax injection equipment does not really come within the scope of what is generally considered a painting robot and is not referred to further in this paper. Six-axis robots are, however, used for externally coating the underbody during the painting process.

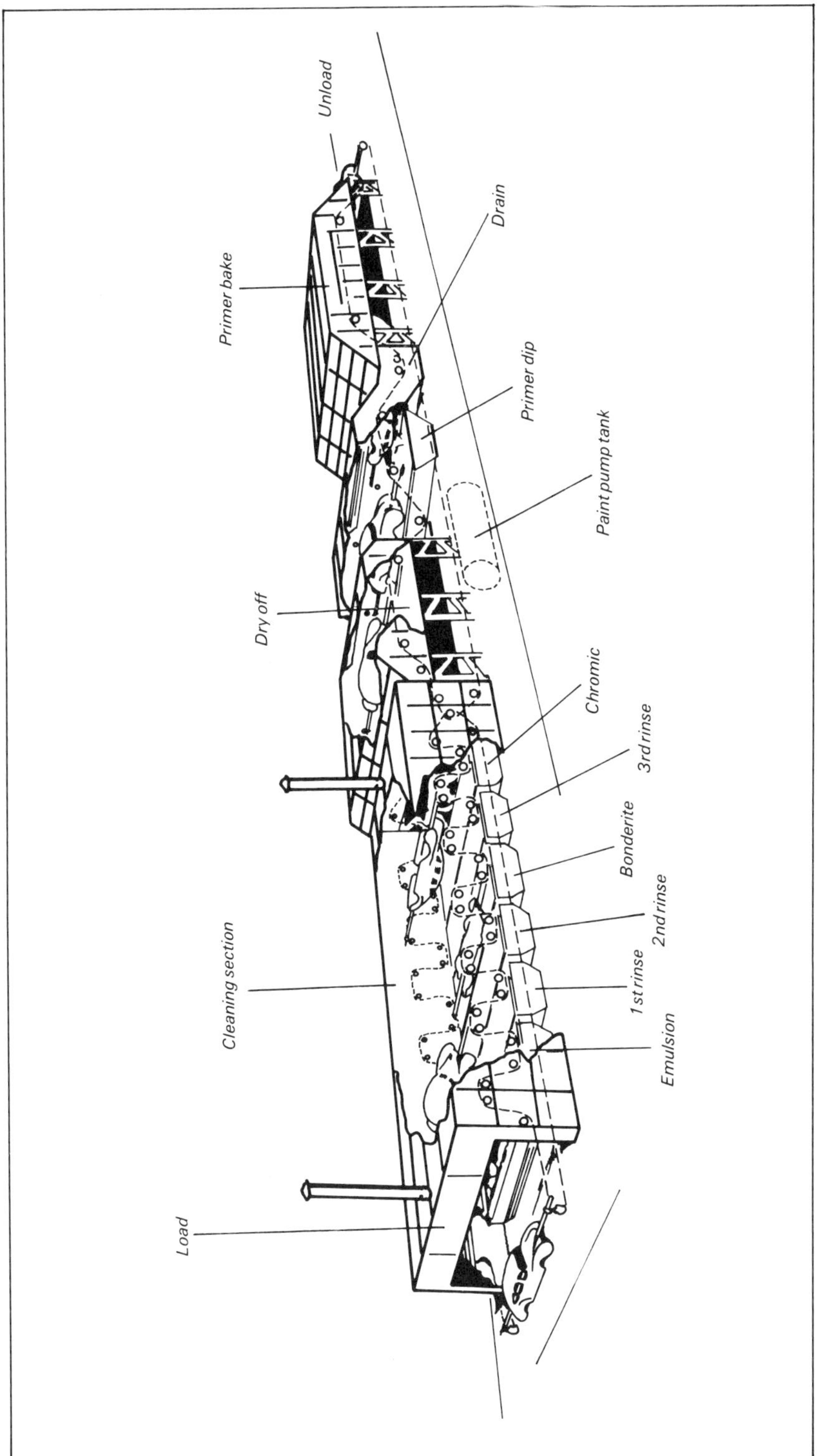

Fig. 1 Automatic pre-treatment and priming plant – 1947

Finish coat application

Following installation of a pilot plant at Chrysler Desotoo Detroit, based on a UK patent, serious attempts were made to spray external coats of paint automatically in the early 1950s at Austin Motors, Longbridge. Machines were installed in all spraybooths to apply fillers, sealers and two coats of finish. Twin sprayguns were mounted in a cam-operated gun arm at a fixed location in each booth whilst the bodies were rotated as they moved past the gun station by the plant conveyor.

Thus, the paint was applied in a helix with the guns being triggered and their attitude (nod) being controlled by cams on a rotating drum. These were what we would now call 3-axis robots (Fig. 2). Whilst many thousands of finished bodies were painted in this manner there were fundamental problems in the concept, not least being the maintenance of an acceptable dynamic balance of the body in its cradle and the effect of any hesitancy in the movement of the conveyor.

These early attempts to apply finish coats were abandoned in favour of reciprocating machines (Fig. 3) or fixed electrostatic applicators of various designs but it was not until 1958 that an installation was commissioned at Morris Motors Cowley, where all external coats of paint were applied automatically on a continuous high-volume production line. Simple 3-axis robots (reciprocating machines) were used in sets of three - two side and one top.

Fig. 2 Rotospray 3-axis robot – 1951

Fig. 3 Finish coat reciprocating machine – 1958

Simultaneously, development work was being carried out in Europe, particularly in West Germany and Italy, on the use of low-speed spinning cup electrostatic machines. These machines and subsequent improvements, including the more recently developed 'turbo' high-speed spinning cup electrostatic applicators, are capable of applying good quality finish coats to approximately 80% of the external surface of a body. Unfortunately, the metal flake of 'metallic paints' is influenced by the process, producing patterns which may be visible in the final finish. There is also a tendency for preferential deposition on edges. This is not the case with reciprocating machines which produce a more even film, although the areas requiring touch-in are somewhat greater.

Current developments

These installations apply paint to the 'easy parts' of the body but there remain small areas requiring touch-up and the difficult window mouldings, door shuts and interior painting.

Multi-axis robots

There is intensive development work being carried out on the painting of these difficult-to-get-at areas, using multi-axis robots, particularly in the USA, Japan, West Germany and the UK.

Painting robot selection

Unlike some other robot applications for advanced manufacturing it is essential that linear and angular velocity of the sprayguns can be programmed. Repeatability of these two parameters is as important if not more important than precise programmability and repeatability of the track of the guns themselves.

It is relatively easy to evaluate the performance of a multi-axis robot in terms of position, repeatability and accuracy. It is not so easy, however, to evaluate repeatability in regard to motion, i.e. path, velocity and acceleration.

It was not until sophisticated motion analysis systems became available that it was obvious that a number of reputable robots performing very well on tasks calling for accuracy and positional repeatability, perform very poorly in regard to motion repeatability and accuracy. These would be quite unsuitable as painting robots for which high (1–1.5m/s) and controllable velocity along with a fairly good path repeatability are essential characteristics.

If these different areas are to be painted with the body on a moving conveyor then all programs must be relative to the body position. This is usually facilitated by incorporating an *X*-axis parallel with the conveyor, additional to the normal 6-axis axis (Fig. 4). Programming is much improved if the movement on this *X*-axis can be programmed, relative to but not necessarily at, the synchronous speed of the

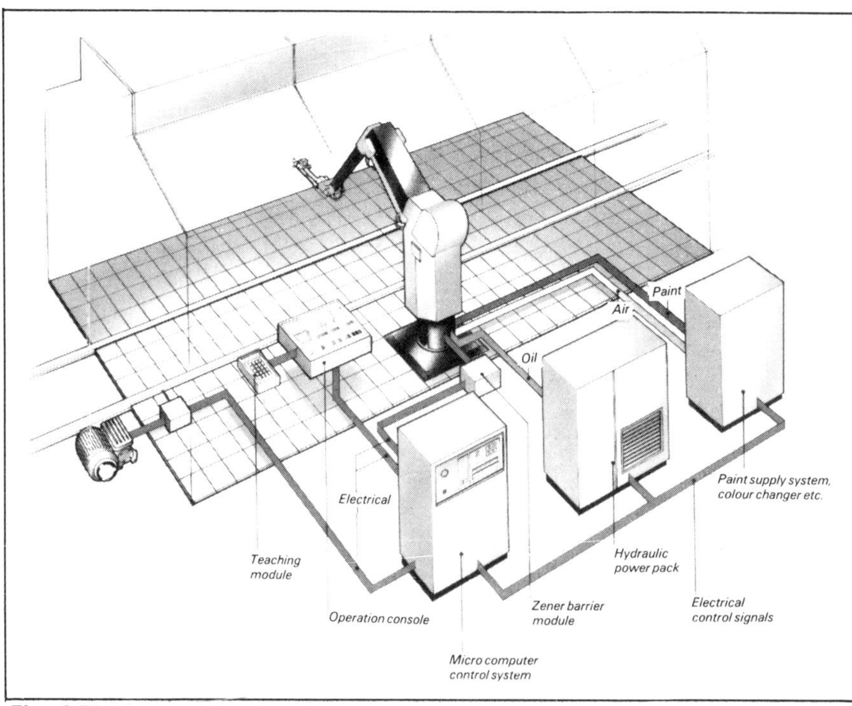

Fig. 4 Rigid armed 7-axis robot

conveyor. In addition, a desirable feature is that the robot will complete each block of its program in the event of conveyor stoppage.

Whilst modern paints do not present the same fire/explosion risk associated with the previous cellulose paints there is a requirement for any electrics within the spraybooth to be flame-proofed or intrinsically safe. For similar reasons hydraulics are usually preferred for motive power although precautions must be taken to minimise the risk of leakage as any oil mist, vapour or aerosol can be disastrous to the painting process. In particular, silicone additives must not be used in any hydraulic circuits inside the spraybooth.

Teaching systems

The most critical feature of a robot to be used for automobile body painting is the conception and design of the control system.

The early multi-axis robots used for painting were really man simulators and accordingly teaching systems were devised which required a skilled paint sprayer to move the robot over the surface to be painted and to transfer that program to the memory of the controlling computer. The success of this method is dependent on the ability of the operator to demonstrate the same skill with the robot as he could with a spray-gun. This was satisfactory for painting relatively simple objects but quite inadequate to obtain the high-quality finish required on an automobile body, let alone cope with the complications of door shuts, etc. This also presupposes that the robot is not required to paint more effectively than the manual operator.

Experience over the years of programming the relatively simple three-axis reciprocators and 'rotary' paint machines had shown that to obtain better results than by hand spray it was essential to analyse the spraygun movement, speed and attitude and distance between gun and body when studying the surface areas to be coated.

The emission rate for the guns should be kept constant for a given production rate.

Essential requirements are to be able to program with the body stationary and build into the control system the capability to compensate for displacement of the body due to conveyor movement register by a pulse generator. In this way the robot movement is synchronised to the body movement millimetre by millimetre from stationary up to design maximum speed. As well as facilitating programming, this system permits the robot to finish painting the particular section within its reach despite a conveyor stoppage.

One essential feature of any painting robot programming system is the ability to edit small sections of the program without destroying adjacent sections and retaining continuity in the final program. As mentioned earlier it is also essential to be able to program the lineal velocity of the gun fulcrum and also the angular velocity of the gun shaft to achieve constant film build at the surface to be painted, remembering that the emission from the gun is constant and fan shaped.

Whilst the knowledge of paint technology is advantageous to the programmer the techniques are different from those of a skilled hand sprayer. It goes without saying that large memory capacity is required, probably of the order of 10 megabytes, as it is not unusual to have to cater for 30 different colours and 15 different body styles in the one production line.

Spraybooth design

Bearing in mind that there is a large potential market for retrofitting robots into existing spraybooths, the physical dimension and geometry of a multi-axis robot is worth consideration. The majority of robots designed for assembly and welding have an elbow which cannot be easily accommodated in a booth designed for hand spraying.

For purpose designed spraybooths this requirement is not essential, but nonetheless a good feature, but, more important, the elimination of the need to allow for line operators enables the selection of a new set of design parameters and operating conditions suited to the process. With hand spraying, the experienced spray operator and/or charge hand can recognise variations and, within limits, compensate accordingly - often subconsciously. The machine is unlikely to be able to do this with the present technology but in every case control of all variables at the optimum level is the essential philosophy for full automation.

It is important to control air movement, psychrometric conditions and the parameters of the services to the paint applicators. In particular, the pressure, temperature, viscosity and solids content of the paint must be adequately controlled at the paint mix house. Tomorrow's automated spraybooth should encompass comprehensive management systems providing monitoring and diagnostic information in the form of colour graphics and printouts where appropriate.

Plant concept and layout

Currently, high-volume body paint shops are based on the concept of one continuous conveyor through the pre-treatment and priming lines on overhead conveyors splitting into a number of parallel finishing lines, each with continuous floor conveyors handling outputs varying between 30 and 45 bodies per hour, corresponding to conveyor speeds of the order of 3.0 and 4.5m/min, respectively. This presents the engineer with the difficult task of designing manipulators to open and close doors, bonnet and boot lids and paint interior surfaces 'on the fly'.

Auto painting 'on the fly'

Some measure of success has been achieved in the USA but at great expense in terms of capital equipment and high operating cost with regard to maintenance, cleaning and paint consumption. In the UK, a pilot installation has been in operation on a production line which achieved the hoped-for economies in operating performance and at a lower capital cost.

All such installations suffer from one important weakness in that failure of any one link in the system is likely to affect total output. On the other hand, the effect on the layout and operation to the rest of the paint shop is minimal and such systems are more easily incorporated into existing facilities. More significant, perhaps, is the fact that with a continuous conveyor layout, tried and proved methods of painting the critical major external areas using electrostatic and/or reciprocating machines can be easily incorporated.

Automatic painting with stationary bodies

In order to ease the problem of painting door shuts and interior surfaces considerable ingenuity has been exercised to develop layouts that provide these difficult areas for painting when the body is stationary.

There are three systems being investigated. The first two are:

- *Modular* - The in-line conveyor is arranged to operate intermittently indexing at high speed between painting stations. In order to have sufficient time at each station the output of this system is low (approximately 14 bodies per hour) and the speed of transfer high - too high to be used in conjunction with more conventional equipment to paint the external surfaces. This system is sometimes referred to as modular and high production is obtained by arranging a number of parallel modules each comprising tack rag, spraybooth, flash off, oven and conveyor system.
- *Cell type systems* - With this the body to be painted is 'shunted' into a booth area containing the necessary multi-axis robots, door manipulators, etc. to carry out the application of one or more complete coats of paint.

The above two concepts (Fig. 5) require the multi-axis robots to paint the major and most visible external surfaces. Quite apart from using

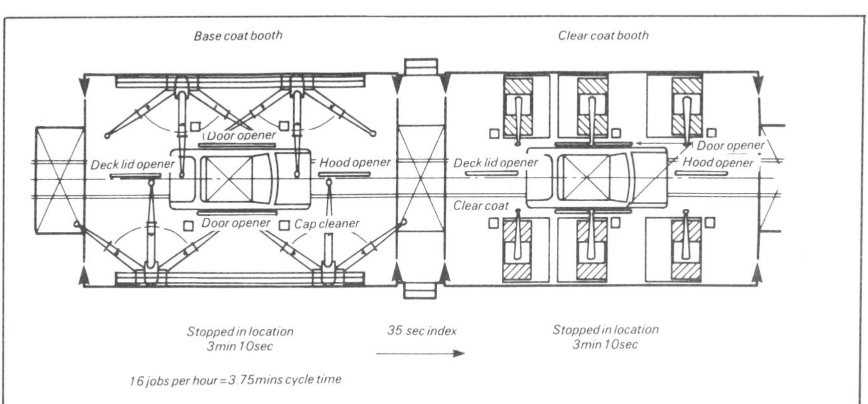

Fig. 5 Modular or cell concept

machines which are unnecessarily complex for the major part of the process, the multi-axis rigid arm robot as available today is not suitable to paint large areas which require a high standard of finish. If air optimising spray applicators are used it is essential to obtain a four-lap coverage which requires a lineal velocity of the guns of the order of 2m/s if the cycle time is to be acceptable. Multi-axis robots currently available are hard pressed to reach 1.5m/s.

Alternatively, turbo or mini-bell electrostatic applicators can be mounted on multi-axis robots. There are the problems due to gyrospcopic forces generated and in matching the finish obtained on the external surfaces with the touch-in and shut areas where it is not feasible to use electrostatic applicators due to Faraday cage effect.

These concepts do have the advantage that in the event of failure of one or more painting systems reduced production can still be maintained, bypassing the defective cell or module.

The other system is:

- *Indexing in-line* – There is a compromise arrangement which combines the advantages of in-line and cell-type concepts but at the expense of complex mechanical handling of the bodies themselves. In this mixed system exterior coats are applied on in-line sections operating at conventional line speeds and using reciprocating or electrostatic automatic machines and touch-in, shuts and interiors painted in cells (Fig. 6).

In terms of overall cost of the capital equipment compared with state-of-the-art standard concepts using electrostatic and/or reciprocating machines for external coating and manual touch-up, door-shuts, bonnet, boot and internals is as follows:

Auto exterior/manual interior	100
Modular indexing	160
Cell	240
Indexing in-line (mixed)	140

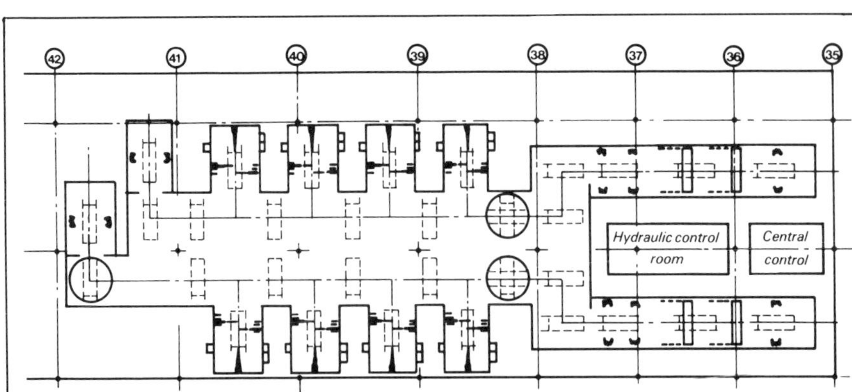

Fig. 6 Mixed cell and in-line concept

Manipulators

No attempt has been made in this paper to cover the problem of opening and closing the doors, bonnet lid and boot lid/tailgate automatically. The requirement is common to all concepts, although the approach will be different. In the case of in-line concepts there are strong arguments in favour of integrating the manipulators with the painting robots and holding the doors or lids in the optimum position throughout each painting cycle. Care must be taken to design the system to minimise overspray build-up on the manipulators and avoid 'dirt' problems.

In the case of cell-type systems many more manipulators are required for a given production but the design can be less complex, being a development of simple 3-axis pick-and-place robots.

Flexible arm robots

The first of a new generation of robot is entering the field which should solve many of the problems with an in-line concept (Fig. 7). It has exceptional reach and flexibility compared to 'conventional' rigid armed multi-axis robots and has a very slim base unit and a 'body roll' feature incorporated in its 7-axis control system. Owing to their long reach and slim base these robots can be arranged at the sides of the booth on fixed bases so that the bodies can pass with the doors open and the robot can still reach inside the body (Fig. 8). Dedicated door and lid openers can be located at the entry end and similar closers at the exit end of the painting zone outside the overspray areas. This arrangement results in a very elegant variation of the in-line concept at slightly lower capital cost - approximately 115% compared to 'standard'. In regard to the risk of total line stoppage in the event of a fault developing with one robot, the dimensions of the robots and the arrangement of the manipulators is

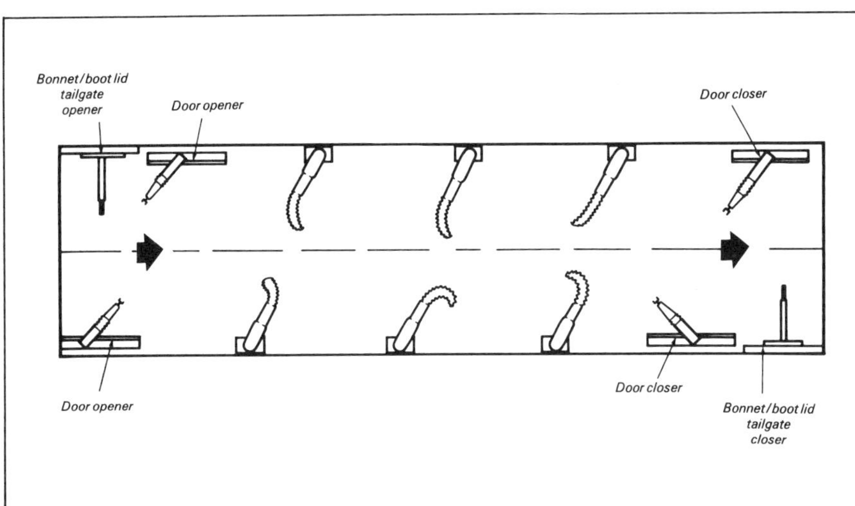

Fig. 7 Interior spraybooth: in-line continuous concept fixed robots

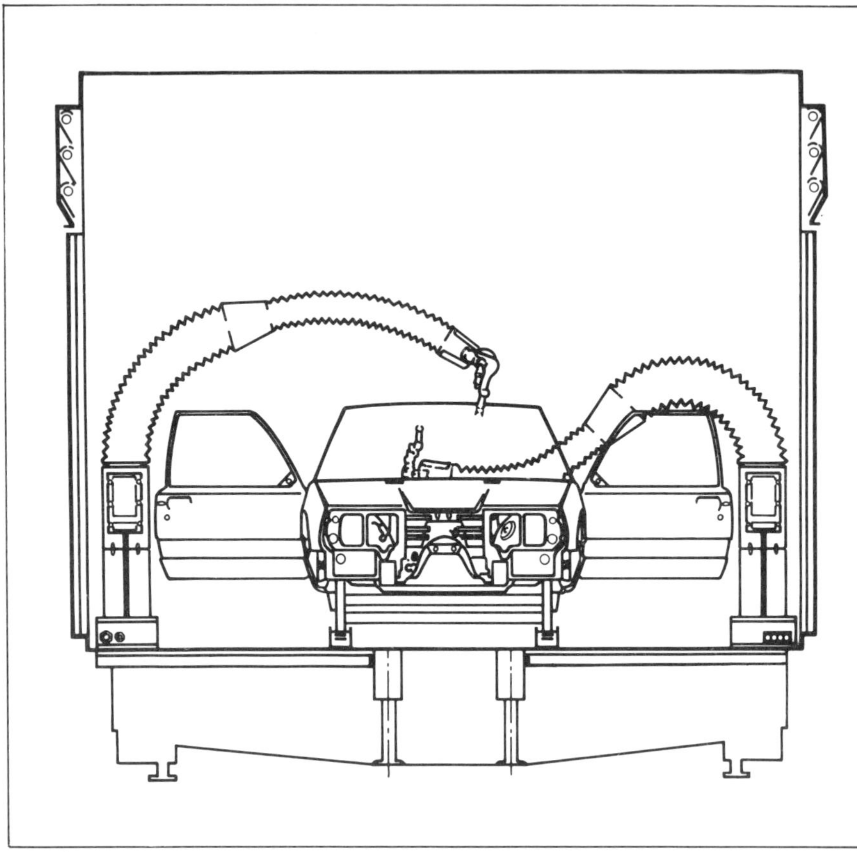

Fig. 8 Fixed base flexible arm multi-axis robots

such that they can all be stowed away on the walls of the booth and leave an unobstructed booth for manual spraying.

Importance of full automation

The advantages of automatically painting the external surfaces of the body are substantial. Thereafter there is little to be gained unless the use of manual operators in the spray booth can be completely eliminated. Partial automation of door shut and interior painting is very costly and unlikely to show any commercial advantage. In contrast, full automation opens up the possibility of using new coating materials and processes and avoids the compromises necessary when an operator is in the booth.

General Motors in the USA is the only manufacturer attempting the automatic painting of the complete passenger car body on a full production scale. Five new plants have been designed around the modular cell concept and are in various stages of completion. The first of these has started production trials. Programming the robots is a very lengthy process but well worthwhile provided reliability and a high standard of finish, particularly of the large external surfaces, can be guaranteed. This would open up new options for the paint technologist and process engineer.

The capital cost of a fully automated plant of this type is very high as is the complexity and sophistication of the control systems.

A new approach to body finishing

All the previous concepts have been based upon painting the body in assembled condition, as is current practice. With the prospect of a change in the materials and construction of the passenger car body, specifically the greater use of 'plastics', now is surely the time to develop a new manufacturing philosophy incorporating all the new advanced manufacturing techniques in the body-in-white, paintshops and trim shops without being influenced by the limitations of systems which have been evolved to suit largely manual operations carried out on a continuously moving or intermittent conveyor.

A separate metal cage or skeleton, incorporating door shut areas, engine bay etc., lends itself to modular construction and painting with indexing or cell-type concepts. The large external surfaces, whether they be of metal of plastic, can be better finished on continuous-in-line systems. Trimming could be carried out on subassemblies which are then brought together towards the end of the assembly line. With modern material handling techniques and the accurate control essential to any fully automatic handling system, the fear of colour matching and assembly damage problems should be eliminated.

With this system, the quality of the finished body would be higher, the percentage of rejects lower and plant down-time reduced. The equipment would be simple rather than complex and is based on 'known art', both fundamental to reliability. A bonus, which should be attractive to the car owner and his insurance company, would be the easy replacement and/or repair of individual body panels.

As so often is the case 'nothing is new' and the basic concept was adopted by the Rover Company at Solihull in 1963 when producing the P6 model. The skeleton was dip primed and the external finish coats applied by spinning cup electrostatic machines to the body panels mounted on floor conveyor trucks. The finish coats were cured in electric radiant ovens with minimum air movement. Unfortunately, this concept was abandoned for the SD1 Rover model for reasons which are not altogether clear. With today's advanced manufacturing techniques in the press and body-in-white shops, improved electrostatic painting systems and the latest final assembly and trim techniques, many of the alleged problems no longer apply. It is, of course, essential that any reappraisal includes the study of the whole philosophy of body design and construction and is not just confined to the paint finishing shop.

4

INSPECTION

With increasing importance being attached to quality, the role of inspection rightly is one that should not be overlooked. The concept of world class manufacturing and just-in-time supply requires high standards of managers and engineers who must build quality into the system at the very earliest stages of concept design. Quality has to be built in – not inspected out. The result is that inspection should have a positive role to play in achieving high standards of excellence in the finished product. Here again new processes and techniques are being evolved with which every production manager should be familiar.

VISUAL INSPECTION AND CALIBRATION TEST BY ROBOT

R. Mills
Gaydon Technology Ltd, UK
(formerly BL Technology Ltd, UK)

This paper describes and discusses an application of both robot and vision to inspect automatically and visually the correct functioning of an automobile instrumentation panel. The paper includes selection criteria for the system equipment together with the approach used to control both the machine vision and the robot from a supervisory microcomputer. The problems encountered during the system development are discussed.

There is a need for more automated inspection to complement the ever-increasing automation within the manufacturing and assembly process. Automated inspection eliminates the subjective element of manual inspection especially where the inspection work content is purely visual appraisal.

The system was developed to demonstrate the ability of an automatic, flexible instrumentation inspection cell which could be readily reprogrammed by a non-vision specialist. The particular instrument panel chosen for the investigation was from an automobile but the findings and problems would be similar for any similar instrumentation not associated with automobiles.

Because the system was a vision and robot based inspection cell which also tested the display of calibrated instruments it was given the mnemonic VICTER - Visual Inspection and Calibration Test by Robot (Fig. 1).

The system uses only one camera mounted on the end of a robot arm to achieve a 700mm range of camera positioning. This flexibility was to accommodate a mix of styles and models of fascia panels.

One of the main aims besides achieving a reliable, non-subjective inspection cell, was to produce a real-time system using equipment requiring only a short learning curve by non-specialists which could be readily integrated into a production line.

Fig. 1 Overall layout of VICTER

Instrumentation panel description

The instruments to be inspected were in-situ in a fascia dash-panel within an area of 600mm × 90mm and had a depth variation of approximately 100mm. These instruments ranged from internally illuminated push buttons, vacuum fluorescent displays and incandescent illuminated warning displays.

In detail these are as follows (Fig. 2):

- Five control buttons. Each with an internal incandescent low-wattage bulb giving a black/white display. These were inspected for correct motif in the correct assembled position and correct functioning of illumination.
- A seven-segment speedometer. This had three numerical characters and was driven by an oscillating signal. The speedometer calibration was tested by checking for the correct display readout for a given frequency input signal.
- A vacuum fluorescent fuel gauge. This was made up of discrete segments, each segment being illuminated in proportion to the fuel present. This was driven by a dc voltage signal. It was inspected for the correct number of energised segments against a specific known dc voltage input.
- A vacuum fluorescent temperature gauge. This was of a similar arrangement and operation to the fuel gauge. Again this display as with the fuel gauge was checked for the correct number of illuminated segments against a dc input.

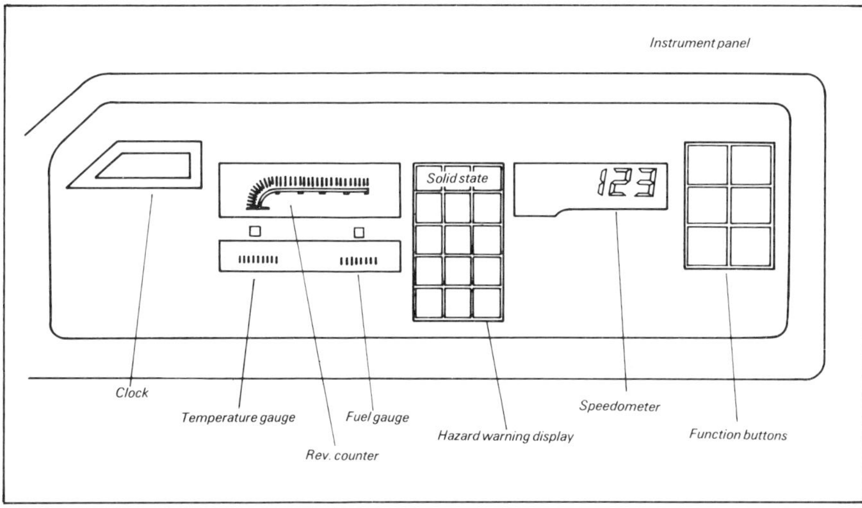

Fig. 2 Typical instrument panel

- A revolution counter. This had vacuum fluorescent illumination and was made up of discrete segments, but unlike the fuel and temperature displays, the rev counter is driven by a varying frequency signal. Its calibration is tested by checking for the correct number of illuminated segments for a known frequency input.
- Twelve condition warning light displays. These were of various colours each with a graphic symbol printed in the centre of the display area. These displays were back illuminated with incandescent bulbs. The displays were checked for correct level of illumination and gross defects in the graphic legend (the resolution of the system was insufficient to detect very small defects).
- A seven-segment digital clock. This was checked for correct time setting.

This gives a total of 22 displays with 47 tests being conducted during the cycle.

These instruments were driven externally from a dc voltage source and frequency signal generator circuit. Each of the 22 instrument panel inputs was switched by a supervisory microcomputer via an interfacing drive card. Whilst it was not necessary for normal cycle operations to switch each device, it was a useful facility in fault finding.

The externally supplied driving signals were physically connected to the instrument panel at two existing standard plug socket connectors in the wiring loom.

General description of system

A solid-state camera with a fixed focal length lens was mounted on the end of a robot arm and the video signal from this camera fed back into a machine vision unit for processing and analysing the image.

The results of the image analysis were communicated along a serial RS232c link to the supervisory microcomputer for further analysis and presentation in the form of a status report issued at the end of an inspection cycle.

There were nine different camera 'shots' or views necessary to cover the 22 displays on the instrument panel.

There were just two operator controls for normal running of inspection cycle. One switch to initiate auto cycle and an emergency stop button.

System equipment

The camera

A solid-state camera was chosen not only for its robustness against knocks and vibration but also because of its compact physical shape as access through the steering wheel was a requirement.

The camera sensor was 320H × 244V. The field of view was 90mm x 67mm giving a resolution pf 0.3mm/pixel. This field of view is determined by the length of the rev counter display. A larger field of view is detrimental to the required resolution.

The camera lens

This was a 35mm fixed focal length, 'C' mount lens which was used at its maximum iris opening stop of 1.7. The reason for using this lens was its low f-stop value.

Because of the fixed focal length of the lens the field of view remained constant throughout the cycle. The depth variation between the various sectors of the instrument panel was accommodated by the robot bringing the image into focus.

The robot

This was a six-axis jointed arm configuration with dc motor drives[1]. The robot was chosen for its ability to manoeuvre the camera within the steering wheel and to hold a steady, non-vibratory position with no positional hunting.

The vision system

This was a proprietary[2] grey-level analysis unit which was chosen for its easy and rapid reprogramming facility and short program software development time. Because the layouts of the various images of the instruments were known, it was not necessary to have the random feature-finding routines which are obtainable from binary vision systems (i.e. routines to find positional values of randomly occurring blobs within the field of view).

This particular vision system also had an image realignment routine to correct for any inconsistency of the instrument panel position or robot position (Fig. 3).

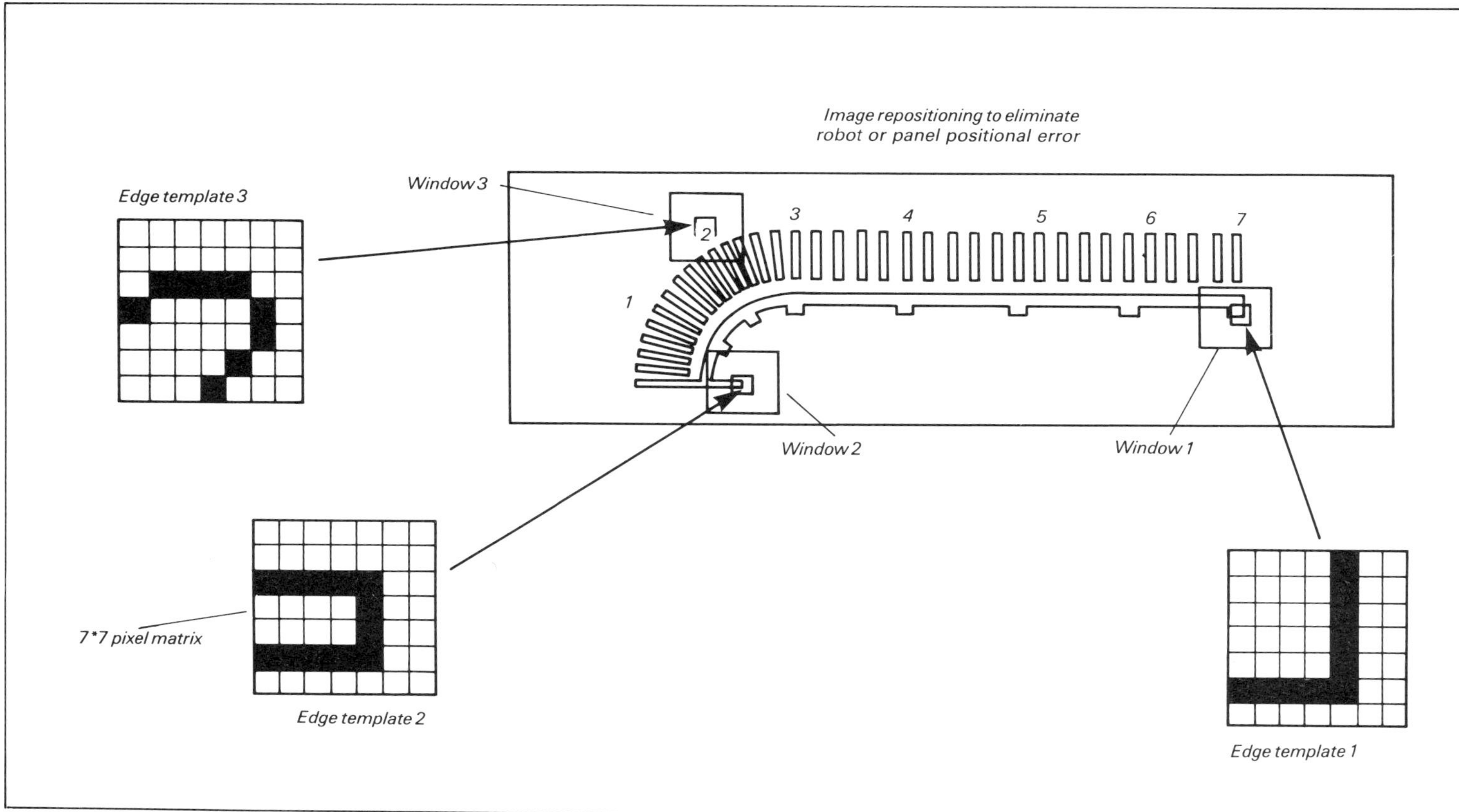

Fig. 3 7 × 7 pixel matrix searches window for correct template match

The information output from the vision system was either of a decimal value or a true or false statement. To obtain the true or false statement the vision system was programmed using a combination of its menu and light pen to compare high and low limits which were set on the data obtained from a satisfactory instrument display.

The programmer decides whether he wants the actual values obtained from the image for subsequent processing outside the vision system or to let the vision system do the processing and to issue a true or false value.

The supervisory controller

This was a standard microcomputer[3] used to undertake status checks on the robot and machine vision unit and to synchronise the operation of these two pieces of equipment.

The microcomputer was in control when the robot went from one point to another and received acknowledgement from the robot controller that each point had indeed been reached. Similarly it was in control when the vision system took a frame of image information and decided which vision program routine to use to process the image. Once the image had been grabbed the vision system issued a 'proceed' signal to the microcomputer.

The microcomputer was also used to analyse the results and give a status report at the end of the inspection cycle. The end of cycle report was in the form of a colour graphic display and whilst this form was not suitable for a trackside hardcopy quality report, it was used with great effect in the early stages of development to show that the system was in fact detecting introduced faults.

The use of a supervisory controller gave great flexibility to the overall system and various sequences of inspection could be obtained by simple changes in the micro's program.

The microcomputer was programmed in BASIC and could deal adequately with processing the data without slowing down the cycle.

Problems encountered

In common with the majority of machine vision applications the problems were to be found in illumination.

The difference in sensitivity of the solid-state sensor to that of the human eye gave the effect of large differences in light levels between the various self-illuminated panel displays as seen by the vision system.

What appeared to the human eye to be a good contrast subject, such as the vacuum fluorescent displays, required maximum iris opening. However with the iris setting the hazard warning lights gave a light-saturated image, completely obliterating the motif.

This problem was in part overcome by using an infrared filter (BG38, 3mm thick) but whilst this overcame the glare from the orange and red lights, the vacuum fluorescent displays and the blue main beam light failed to give sufficient contrast.

This was overcome by utilising the video gain potentiometer within the camera. By hardwiring from the potentiometer to an external junction box an automatic switch was set up, controlled by the microcomputer, to achieve different video gains at the required points in the inspection cycle.

An alternative to this would have been to use an EE (motorised iris) lens but this would have added to the cycle time.

Cycle time

The overall cycle time was 35s which was made up of 15s robot movement and the remaining 20s for the vision analysis of the 22 displays and the 47 indication tests together with the image re-aligning routines which took into account any differences in the position of the instrument panel from that which was used during the original programming.

The graphic display was an additional 25s to the cycle time. This form of status reporting was only used to demonstrate that the system was indeed functioning correctly. A far speedier and more permanent form of reporting is required such as a hardcopy record of failed elements only.

System test routine

To test the system a fault generation routine could be called up from the supervisory controller program. This routine used a random number sequence, each number activating certain switches on the panel driver interface card resulting in a particular instrument fault.

Areas of application

It would be proposed that the system could be used at the instrumentation subassembly stage prior to fitting into a vehicle thus allowing easier rectification. Alternatively it would be possible to inspect in-situ, at the final assembly stage. Although fault rectification would be more awkward at this stage it would have the advantage of being a quality test of the panel on its final assembly position.

A further advantage of using this system at the final assembly stage would be the opportunity to link up with the currently existing electrical test equipment. This would not only sense the amount of current drawn through the operated piece of electrical equipment, but the visual inspection system would check the correct functioning of the associated indication display.

A further increase in the value of this cell would be a development of the system that could use the robot to physically operate push buttons, steering column switches and heater controls checking for correct operating forces.

Limitations of the system

As with most machine vision based systems the results are affected by ambient lighting conditions, but not critically so. Provided the general light level is not high then reasonable light fluctuations are acceptable.

The particular vision system used in these trials worked completely from grey-level information and applied gradient finding routines to determine the edges of features within the image (Fig. 4). A fluctuation in light level does not change the relative edge gradient of that feature.

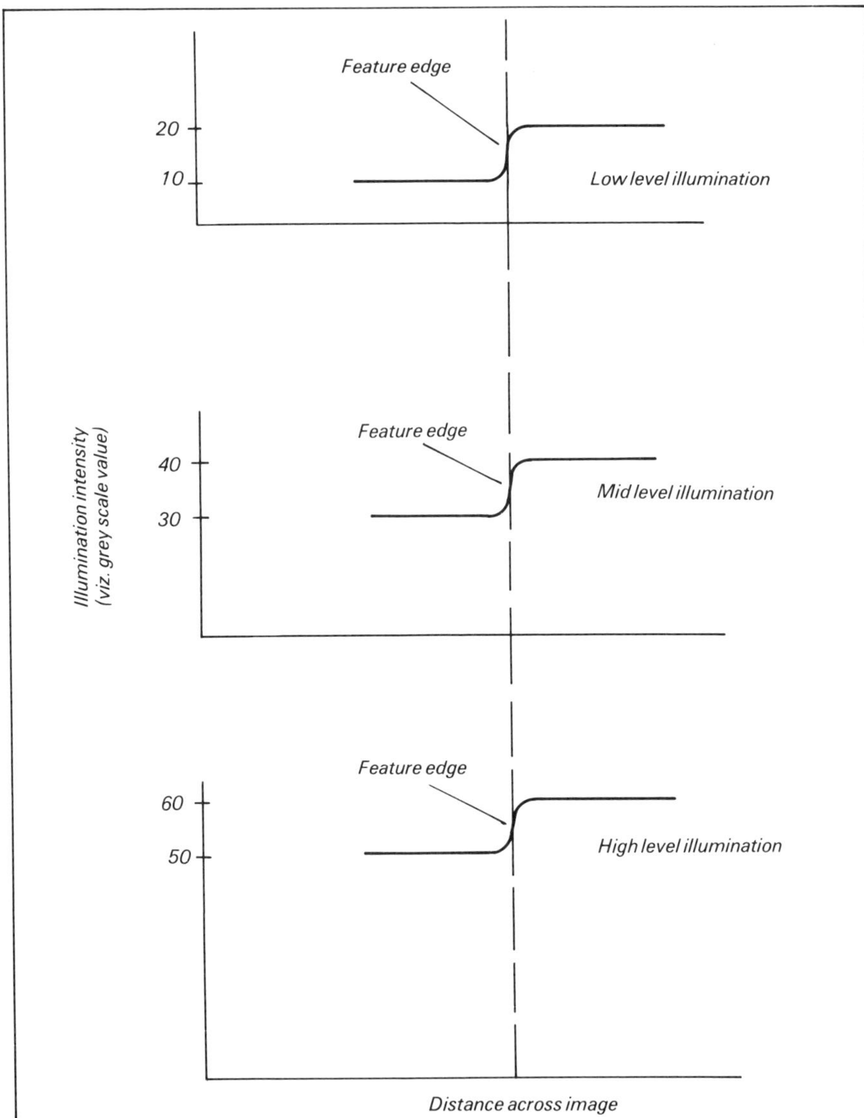

Fig. 4 Using grey-scale analysis: position of feature edge remains approximately constant when illumination changes (within limits)

Concluding remarks

The system was developed to demonstrate the ability of an automatic, flexible instrumentation inspection cell which could be readily reprogrammed by a non-vision specialist. It is considered essential that such an inspection system, which would have to deal with design modifications and new display styles, should allow non-specialist reprogramming and 'tweeking'. Robot reprogramming is now commonly performed by maintenance personnel, and the associated vision system must also allow non-specialist reprogramming (Figs. 5 and 6).

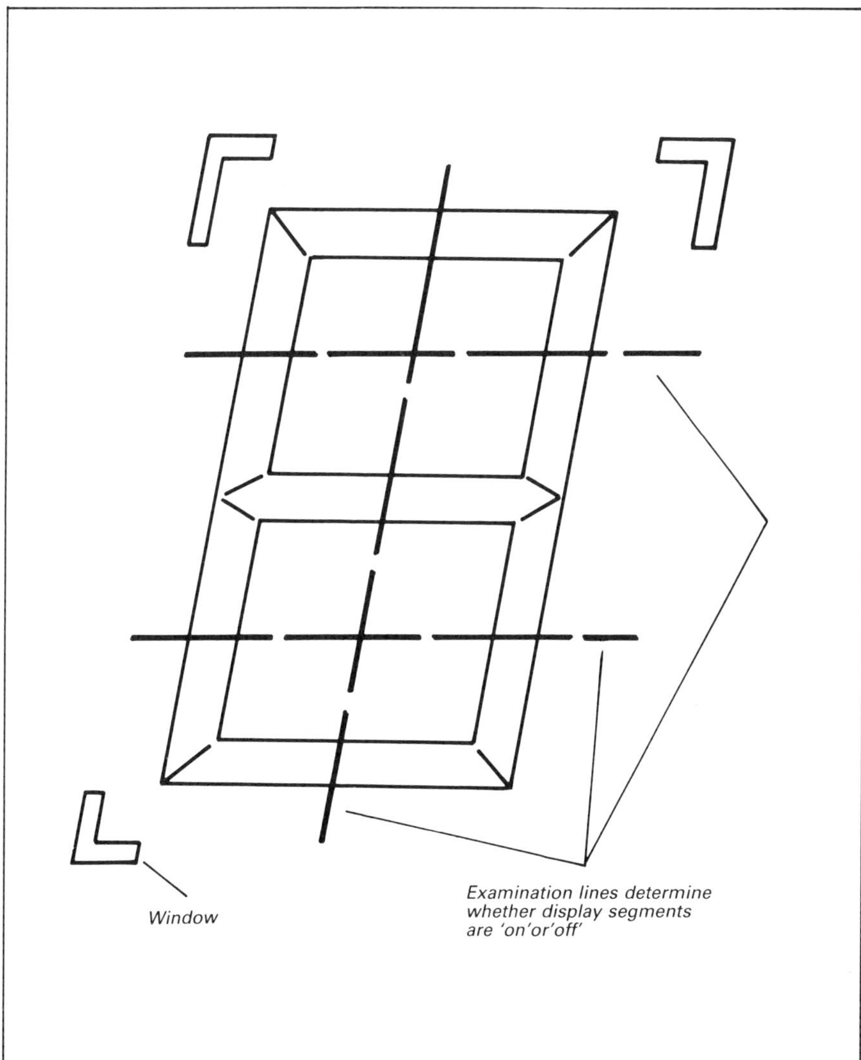

Fig. 5 Seven segment display analysis

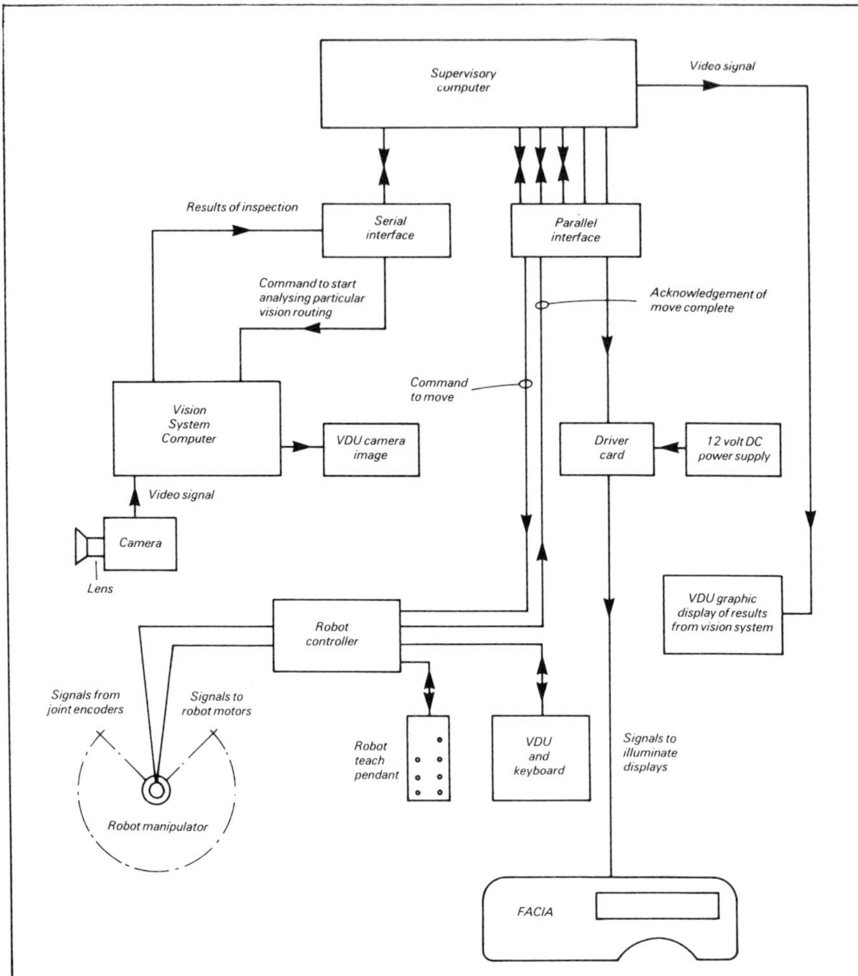

Fig. 6 Schematic of VICTER system

It has been shown that a wide variation in panel illumination was accommodated, and the 22 displays taking 47 individual vision tests to analyse them could be inspected within a 35s cycle.

References

[1] Reflex Robot Operators and Users Guide Rev 2.1 April 1984. Rediffusion Robot Systems Limited, Manor Royal, Crawley, W Sussex.
[2] Itran 8000 Operating and Programming Manual. Itran International Ltd, UK.
[3] IBM Personal Computer XT Technical Reference. IBM UK International Products Ltd.

VISION AIDED INSPECTION TECHNIQUES FOR THE AUTOMOTIVE INDUSTRY

D. German
Ford Motor Company, USA

Renewed importance has been placed on high-speed measurement techniques in light of statistical process control. Advances in measurement techniques will be discussed in three main categories: (1) high-speed matrix camera vision systems installed in various Ford Motor Company plants measuring fasteners 100% for transmission and chassis engine and assembly; (2) laser vision systems for non-contact measurement of surface roughness on transmission components; (3) advanced flaw analysis on transmission and chassis components.

The transmission and chassis division has pioneered the application and development of advanced technology to solve inspection problems. These problems could not be handled by conventional methods due to inaccessible geometry and/or speed requirements. In many cases, the requirements have been satisfied by the development of new technologies or the adaption of technologies developed for laboratory implementation for suitable operation on the plant floor environment. In many cases this has required close cooperation and development efforts between the Ford Motor Company and its suppliers of high technology equipment.

To date the installation of 30 systems reflects the use of a broad range of sensing technology including laser, optic, ultrasonic, nuclear and capacitive techniques. This paper concentrates on the use of laser and optic or vision techniques and how they have been implemented to solve plant floor inspection applications.

Automatic fastener inspection

Current transmissions utilise a large number of fasteners particularly bolts. In the current age of automated assembly, fully automated insertion and torquing techniques are predominant. Thus fastener quality becomes a major concern.

Some 30% of Ford's torque problems are generated by incorrect or damaged fasteners. Additionally, defective fasteners such as components with damaged integral washers, incomplete, unthreaded or completely mixed parts can often jam automated feeding equipment delivering these fasteners to the automated assembly station. These jams are time consuming and may require the shutdown of a complete automated assembly line until the jams can be cleared.

A survey taken of purchased fasteners has shown that an average of 1–3% of all purchased fasteners contain defects. Removing these defects from the assembly process can reduce warranty and in-house repairs and improve productivity on automated assembly lines.

In 1980 two optical fastener gauges were used in a pilot programme at Ford's Livonia and Sharonville plants. The $20,000 systems, one laser based the other filament projection, are both capable of inspecting at a rate of 15,000 parts per hour. Although both systems have proven cost effective, they are limited in adjustment for range and criteria requiring complex procedures to change between part numbers and typically can inspect only one to two variables. Flexibility in such inspection systems is highly desirable since the inspection part rates far exceed the fastener requirements for a given line. Therefore, flexibility to rapidly change over between part numbers, to use a single system as a receiving tool inspecting all incoming parts in batches, is highly desirable and very cost effective.

At Ford's request, Diffracto Ltd, a Canadian gauge supplier, has refined the optical matrix concept. The concept is to establish an *X-Y* grid of microscopic picture elements of pixels utilising a solid-state matrix camera. Virtually anything that is placed within the range of the camera can be inspected without mechanical adjustment. Such a matrix camera system has been installed at the Livonia transmission plant. It has been programmed to inspect 30 different fasteners which is the total range used in both the AOD and the C6 transmission. Up to 11 different parameters (Fig. 1) can be inspected. Each parameter is compared to

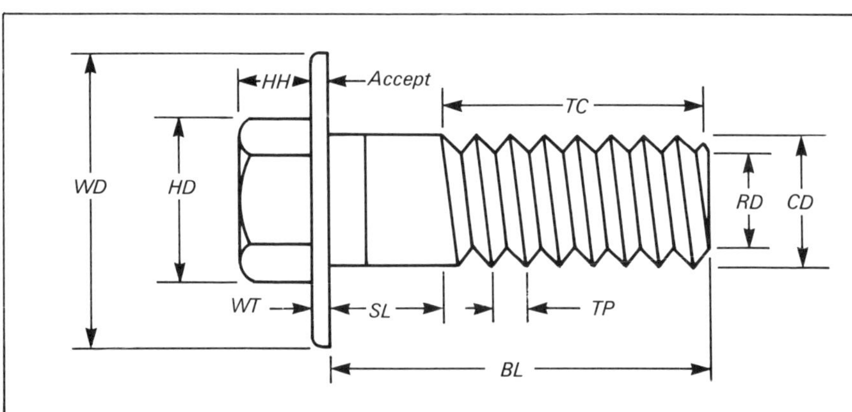

Fig. 1 Fastener dimensions measured

high/low adjustable tolerance limits. The system has the capability of storing a full set of parameters for up to 170 different fasteners.

The system consists of a vibratory bowlfeeder with adjustable tooling, an inspection station which utilises a stroboscopic light source to inspect parts on the fly, an automated sorting gate to remove parts outside of tolerance limits from the batch, and a microprocessor control unit which also maintains statistical information for later analysis and can provide hard copy data to be used in supplier quality surveillance. A diagram of the system is shown in Fig. 2.

The $40,000 investment in this piece of equipment is expected to yield annual savings of $250,000 in labour, repair and warranty.

Non-contact surface roughness measurement

A number of components in current automotive transmissions require careful control of microfinish for proper operation and long life of the transmission. In a number of cases, the measurement of microfinish on transmission components has been implemented through the use of laser-based sensors.

A typical example of such high technology problem solving can be seen on the AOD output shaft. In the manufacturing process, the shaft is machined on high-speed lathes. Carbide cutting inserts generate grooves which will receive mating cast iron seals. Compatibility of the seal with the groove is a function of the surface finish of the groove side wall. Seal life is critical to governor and direct-drive operation and may result in total transmission failure. If appropriate microfinish was not achieved during the AOD launch, a surface finish problem on the output shaft ring groove was discovered. All production shafts were

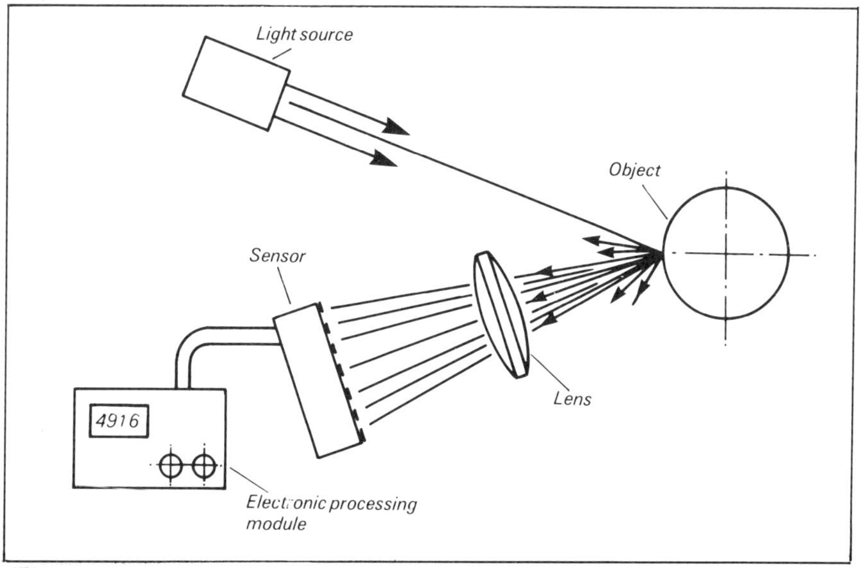

Fig. 2 Laser surf: theory of operation of surface measurement

suspect, therefore, a 100% profilcorder inspection was ordered. This is a contact or stylus type technique. The inspection, however, proved to be both time consuming, requiring approximately 30 minutes per shaft, and inaccurate as geometry allowed only a single-point contact.

Since such inspection procedures were totally impractical for use in volume production, a vision system supplier was contracted to provide a prototype laser system. This was developed in nine days. In operation, a laser beam is directed at the surface to be inspected and reflects from the surface onto voltage sensitive receivers which analyse the light-scattering envelope from the surface. This envelope is correlated in the machine controller to surface finish. The part is rotated 360° in the inspection zone to determine the worst case microfinish on the surface. All laser readings were correlated to test end and road test failure.

No field failures have been reported since the gauge has been in use. Following the success of the prototype, two semi-automated systems were acquired and are currently in daily production use. These production gauges have a total cycle time of 5s and inspect both sides of eight sealed grooves. Although the production gauges were intended primarily to monitor quality, it has been determined that feedback information is also available which is used to control the feed rate of the lathes and tool and coolant change frequency. The $200,000 investment for this project has resulted in a cost avoidance of almost $3 million per year.

Advanced flaw analysis

Inspection of components for surface defects has traditionally been carried out by visual inspection methods. That is, inspectors are placed at appropriate locations in the production line to carry out traditional

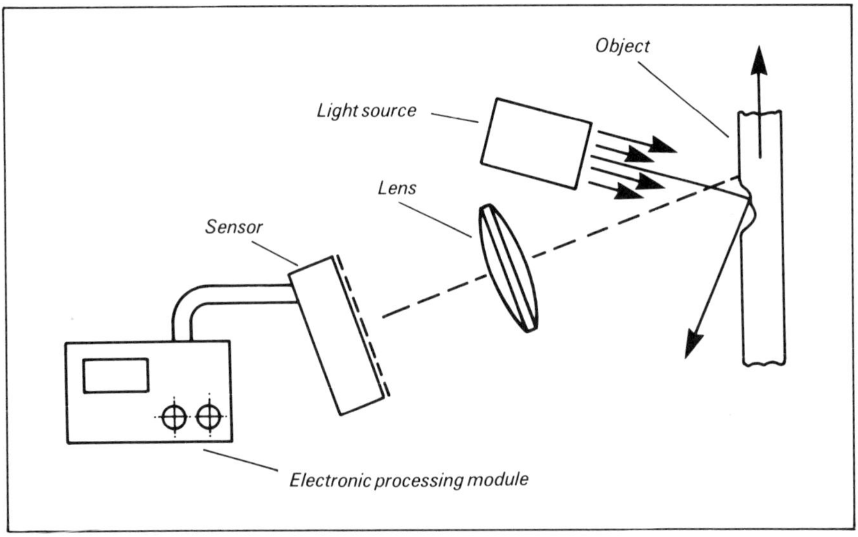

Fig. 3 The theory of flaw detection

visual inspection. This approach is not only expensive because of labour costs and related overheads, but also is not totally reliable, a situation which is unacceptable for critical safety components.

Therefore, the transmission and chassis division has actively pursued the use of automated flaw detection equipment. A good example of this is the power steering rack housing. This assembly constitutes the actual power steering mechanism. The piston bore, where the power steering piston is actuated, is a critical safety surface. This component, which has previously been made of cast iron or of aluminium with a steel sleeve, had not presented problems. However, to reduce weight and cost, it was decided to convert to an all aluminium power steering rack housing. This created a significant inspection problem since the bore, approximately 1.5in. in diameter and 16in. in length would be subject to random porosity from the aluminium die casting procedure. A full 100% visual inspection would require several inspectors because of the part cycle rate, which was one part every 8s. In addition, the inspectors would require auxiliary viewing aids such as sight pipes or fibre optics because of the small/large aspect ratio bore. This approach was considered unacceptable both from a cost-effectiveness and an efficiency-of-inspection point of view.

The solution to this problem was found in a semi-automated vision inspection system, two of which have now been installed and operational for several years. Each inspection system uses an automated borescope approach. A specialised probe is automatically indexed down the bore having approximately 0.125in. clearance between the probe and the part sidewall. This probe is used to remote the image back to a solid-state digital image sensor which converts the reflected light image to grey-scale information. This is decoded in a microprocessor control unit which provides signal analysis and accept/reject indication based on programmable limits on defect size. For example, the unit can be programmed to ignore defects under a specified size, as well as providing an area criteria resulting in the capability of rejecting on small grouped localised porosity. A storage display map output is included with the system which is very useful for set-up and visualisation.

This system has proven very successful and has eliminated a potential field safety problem.

Concluding remarks

These examples are only a few of a spectrum of the 30 installations of advanced technology in Ford Motor Company. The application of new technology has solved a number of inspection problems.

OPTICAL GAUGING OF ENGINE CRANKSHAFTS

D.W. Whitney
General Motors Corp., USA

Optical non-contact gauging is used by Pontiac Motor Division for final inspection of four cylinder engine crankshafts. Light is projected past the crankshaft onto a diode array, which determines how much light is blocked out. This shadow is compared to a master shadow, from which scale factors are used to calculate the dimensional measurement. Compensation for temperature differentials between the part and master is performed. Mastering is accomplished automatically. Microprocessors record and analyse 10,000 individual measurements to monitor 54 characteristics for each crankshaft. Accuracy and repeatability of this vision system have been statistically demonstrated to be 0.3μm. The system has been integrated into an existing manufacturing process with no special environmental considerations.

The introduction of interchangeable parts by Eli Whitney caused an immediate need for the means to compare those parts to a standard. Advances to meet these needs have occurred in step with demands to improve part precision and accuracy. Micrometers, calipers, and snap gauges gave way to combination air-electronic and pure electronic gauge systems.

These gauges all rely on contact to perform their measurement. This contact also creates trouble for the gauge. Wear of the gauge head occurs; part geometry variations reduce measurement accuracy. Part wear or damage can occur if contact pressure is excessive. If contact area is increased to lessen pressure, surface characteristics may cause more variation. Compressed air blowing on the part carries with it moisture and/or oil, attracting dirt. Part temperature is also affected by this air; expansion and contraction of both the part and the gauge must be considered. Contaminants, like dirt, tend to jam linkages, and block air orifices, resulting in frequent maintenance. These contaminants may even be detrimental to the part itself.

The opposite of contact gauging is non-contact gauging. The effects of the gauge on the part are obviously gone. Similarly, the part cannot wear or contaminate the gauge. When dealing with desired measuring accuracy and precision (repeatability) at levels of 1μm or less, the attributes of non-contract gauging are most attractive.

A type of non-contact gauging is the venerable optical comparator ('shadowgraph'). Its drawback has been lack of repeatability among operators and slow usage speed; its advantage is the ability to measure a wide variety of dimensions without special fixture requirements. A shadowgraph operator, assigned to measure parts on a typical mass production schedule, would soon fall behind, make errors, have a nervous breakdown, or, most likely, all three! By adding light sensors to read data and a computer to manage that data, this shadowgraph evolved into a crankshaft inspection system capable of measuring and tracing 54 separate characteristics on each crankshaft produced.

Optical gauging system

The gauging system is completely automatic. Just prior to entering the gauging stations, part temperature is measured. The first gauging station rotates the part to bore, and threaded holes. The second station gauges the part in a stationary mode to provide rod pin journal ovality data. Parts gauged reach the operator station, where any defects found are displayed on a cathode ray tube screen for operator attention. The defective parts are segregated from acceptable parts at this point. For the purposes of discussion, the system is divided into several parts: gauge sensors and light source; characteristics monitored; and data and system management. Fig. 1 shows the system layout.

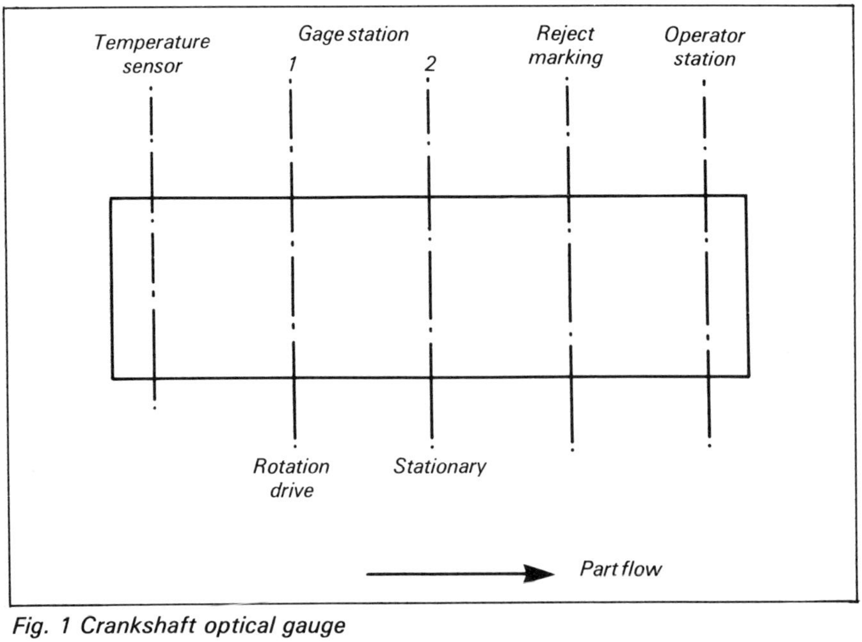

Fig. 1 Crankshaft optical gauge

Gauge sensors and light source

The earlier description of this system as a computerised shadowgraph is fairly accurate, although an oversimplification. A shadowgraph projects light past an object. The shadow cast by the object is focused on a screen at a selected magnification. Measurements are made by comparing the particular feature to a grid on the screen.

For this crankshaft gauging system, light is projected past a part. The shadow cast is focused on a diode array, a device for converting light presence to an electrical output. The diode array, through its arrangement as a grid system, counts how much light is blocked out by the part. The count is then compared to the master shadow count. The unit comprised of the diode array and the focusing lens is referred to as a sensor unit.

The light source is placed directly opposite each sensor unit. A simple incandescent light bulb is utilised; positioning and focus are critical to project the required light to the sensor. It should be noted that with careful alignment of the sensor unit, light source and part, the resultant measurements are along line-of-sight or true tangents, eliminating error due to geometrical properties (Fig. 2).

Characteristics monitored

The crankshaft is probably the most complex machined part in an automotive engine (rotary engines excepted). Added to the complexity of shape, is the need to fit bearing clearances and to maintain piston stroke and timing in the engine assembly. These needs result in the large number of characteristics being monitored (Fig. 3).

A partial list of these characteristics includes: main bearing journal diameter; rod pin bearing journal diameter; pilot bushing bore diameter; oil seal diameter; and gear fit diameter.

The above characteristics are measured directly by the sensor units. Bearing journal taper and ovality, main bearing journal alignment, crankpin/keyway index, pilot bore runout, and oil seal runout are

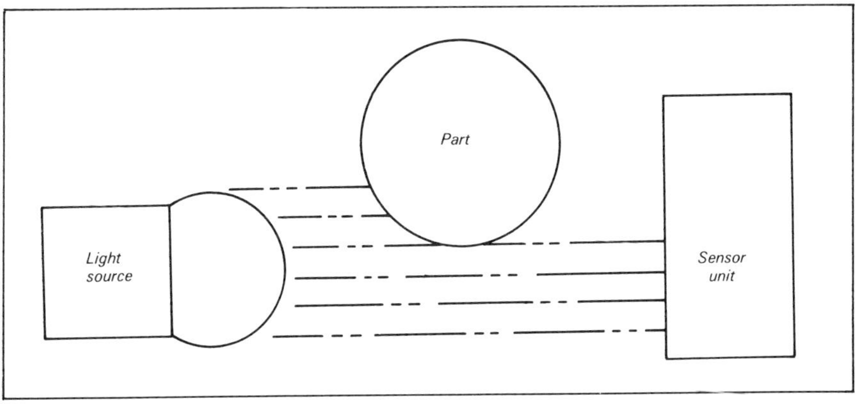

Fig. 2 Sensor unit and light source

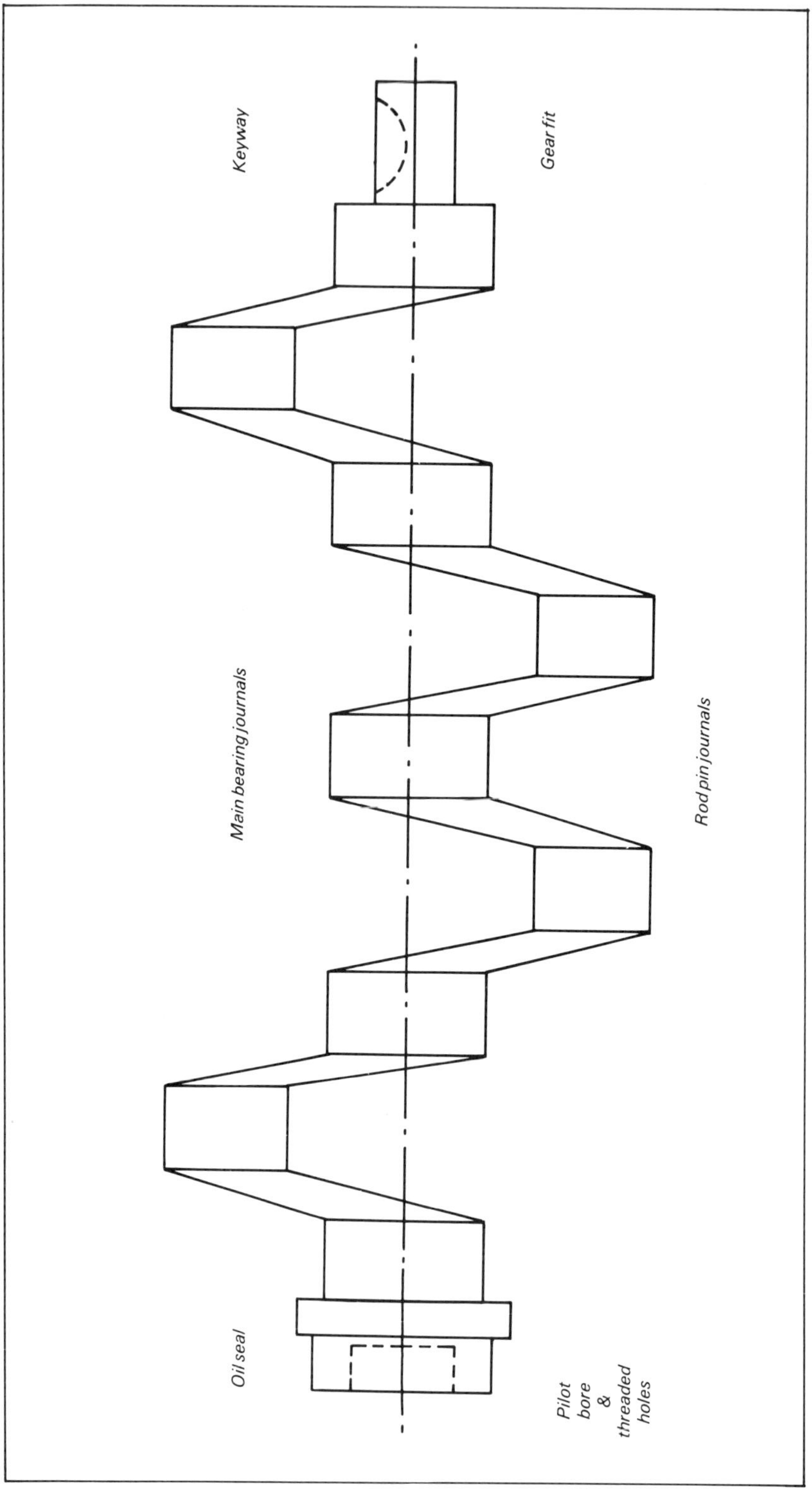

Fig. 3 Characteristics monitored

among the characteristics requiring part rotation and/or calculations in order to be checked. A computer system was married to the sensor units to perform this task. In all, 54 characteristics are checked and monitored by the system.

Data and system management

The computer system devised consists of a master microprocessor and four slaves. Each microprocessor performs specific tasks related to various measurements. The slaves communicate with the master unit, which, along with its own unique measuring tasks, compiles and reports the data as a measurement record.

In conjunction with the above units, a Gould Modicon programmable controller controls physical movements of the parts within the gauge. These movements include station-to-station transfer, part rotation, rotation drive synchronisation, and reject marking. Part rotation and rotation drive synchronisation are required only in the first gauging station.

This data management system performs the task of making over 10,000 individual measurements in order to evaluate each crankshaft.

Measurement dynamics

The development of the data management system has resulted in unique measurement evaluation. Many automatic gauges accept/reject parts based on measurements taken only in one plane, or, if rotated, based on passing over limit lines. Any dirt present on the part which could affect the measurement is ignored. Human operators, on the other hand, average their findings by checking dimensions in several places. The system used was developed to reflect human habit and to take into account the presence of dirt or other contaminants. A great advantage of this averaging is to ensure that characteristics are evaluated equally from part to part. Measurements which do not meet specifications are always noted for the operator; the coding of the defects also allows the operator to evaluate the seriousness of the defect.

Five sections will be used to briefly explain the measurement evaluation and calculation: main bearing journal, oil seal, and gear fit; rod pin bearing journal; pilot bushing bore and threaded holes; temperature variation; data management.

Main bearing journal, oil seal, and gear fit

As the crankshaft is rotated, five flash sensor readings are taken at every 10° of rotation, beginning with the top-dead-centre position of the No. 1 rod pin. The average, in raw counts, is recorded for each interval. Following completion of one revolution, the raw count data is filtered by comparing each record with the proceeding and following record. If the record in question differs from its neighbours by a value exceeding a specified window, the preceeding value is substituted. In this way, dirt, dust, and other contaminants have no effect on the measurement. Once filtering is completed, the data is converted to linear measurements using preset scale factors.

Rod pin bearing journal

Rod pin bearing journal measurement requires special treatment because the offset of the journals from the centre-line moves the journal in and out of the view of the sensors. In order to measure the journals, flash sensor readings are taken only as the pins pass through their respective top-dead-centre position. As above, the raw counts are averaged and filtered. Scale factors again convert the raw counts to linear measurements. A second gauging station is needed to supply data for calculating pin journal ovality. This section takes flash readings in a stationary mode. The averages of these measurements are similarly converted to linear measurements.

Pilot bushing bore and threaded holes

The measurement of the pilot bushing bore and the threaded holes requires different hardware, although the light sensing technique is the same as previously explained. Carbide inserts attached to lever arms follow the bore sides. These arms pivot at midpoint, allowing the other end to reflect the movement of the inserts. This reflected movement is read by the light sensor unit; again filtering and conversion to linear measurements are performed.

The threaded holes are checked for thread depth and thread form. Light is projected into the hole. The amount and pattern of the light reflected back to a sensor unit determines the measurement. The presence of any contaminants is so critical to this measurement that the limits used were modified so that only presence of threads and hole depth are effectively measured.

Temperature variation

Part temperature variations can account for a significant portion of measurement variances. In the load station immediately preceeding the first gauging station, a transducer senses part temperature. This temperature is compared to that of the gauge master at the last calibration. Scale factors are used to convert the measured data to account for expansion of contraction.

Sensor focus can be adversely affected also by temperature. Contraction/expansion of the sensor unit mount varies the focus of the sensor unit with respect to the part. Such movement is prevented by maintaining a constant temperature in the sensor box. In this case, a heater is used to maintain temperature at 40°C (± 1°). An added benefit is the increased life of electronic components which do not experience cyclic temperature variations.

Data management

Once all the diameter measurements are taken, the data management system takes over. Average diameter, taper, and ovality are calculated as needed. In addition, for each bearing journal, an overall average of the two diameters measured is found. The results are compared to the preset limits. If any resultants fall outside the limits, the offending characteristic value is displayed along with a reject code to indicate reason and severity. Main bearing journal, oil seal, gear fit, and pilot bushing bore runout is calculated relative to the centre-line of the end main bearing journals. Ovality of the rod pin bearing journals is calculated by comparing the average values found in both gauge stations.

These characteristic values of each crankshaft are all stored in computer memory until reaching the operator station. At that point, defective characteristics are displayed for the operator. The entire measurement display or portions of raw data can also be off-loaded to a recording device if desired. This feature is very useful during run-off and during quality checks of the manufacturing process. Work which can take a week of painstaking manual activity can be accomplished in an hour.

Another important use of the computer system is for self-diagnosis and troubleshooting of the gauge system. An understanding of the measuring sequence and observation of the data will pinpoint trouble spots in a minimal time period.

Performance

This crankshaft optical gauging system is in use on the plant floor as an integral part of the production process. The first step in this integration was to pass a run-off, both on the vendor's floor and on the plant floor.

In the case of gauges, primary consideration is given to repeatability (precision) and accuracy. These parameters must be statistically stable over time. Pontiac Motor Division manufacturing engineering standards specifies a sequence of measurements in a given time period using approved masters to determine gauge acceptability. Even more stringent tests were undertaken for this system because of the newness of the technology of this application:

- Five gauging cycles for each of the three masters (minimum, mean, maximum) for both mains and pins.
- Five production samples gauged five times each.
- Repeat steps one and two above at 30 minute intervals over four hours (total nine repetitions).
- Acceptability based on data stability, repeatability within 10% of the tolerance band width, and accuracy within ±5% of the tolerance band width from the known value.

These criteria are checked by using X-bar/R control charts and 6-sigma capability. Finding these values is important, since they establish the amount of gauge drift over time, the amount of variance in the measurement due to the gauge itself, and that the variance that is present is predictable.

The repeatability of this gauge was found to be less than 0.5μm. Accuracy was found to be 0.3μm. These values are only 1% to 2% of the tolerance band widths. Even more remarkable is that through these tests involving nearly 20,000 characteristic measurements, only six individual thread measurements exceeded the established limits. A portion of the test was repeated after eight months of system usage; identical results were found.

Production rate of the gauge itself is around 250 parts/hour (net rate is somewhat less due to shuttle system constraints). Calibration is needed only weekly; this interval was found by observing master drift over time. Calibration is performed automatically following loading of the appropriate master. Essentially, this calibration is a gauging cycle run so that the computer can memorise the master shadow and relate this shadow to the master's known value. Only the mean masters are required to perform the calibration.

Maintenance consists mostly of cleaning the glass shielding for the lights and sensors. At one time, this cleaning was thought to be a potential headache; experience has shown that cleaning is usually needed only weekly. Without the oil/water contamination found in most compressed air systems, dirt does not accumulate very fast. The precision part rollers are cleaned and checked periodically. Replacement has been needed only once following a 'wrekcident'. Many cycle safeguards have been built into the system to minimise these occurrences.

Some failures of sensor units have been encountered since installation. The sensors that failed had all been in use during the entire development period (almost three years). Many electronic components fail in industry in much shorter time frames. The light sources have proven to be surprisingly durable. After a full year of operation, the light bulbs were replaced; even at that point, no bulb had failed, although some were a little dim. At various times, sensor sensitivity was adjusted to account for this light variation, enabling the continued function of the gauge.

The system is installed on the plant floor. There is no special floor vibration isolation; no controlled environment enclosure; and no special washer for parts cleaning. Employees have readily adapted to the system operation, giving them more opportunity to track process quality trends and to solve quality problems which may occur.

Concluding remarks

The system described is an operational vision gauge. Many advantages have been pointed out: non-contact measurement; computer-driven to allow management of the enormous amount of data in real-time; self-diagnostics; measurement variation is nil (less than 0.5μm repeatability, 0.3μm accuracy); and minimal calibration drift.

Disadvantages exist compared to conventional gauging already in place:

- Considerable training is required to develop understanding of the system. Sceptics like to see the measurement on a column; they believe they are able to visualise the measurement itself as it takes place and are thus verifying its existence. This system only shows the final result as a linear measurement. It takes some length of time to convert these people.
- Considerable training is required to develop in-house service. Along with knowledge, real interest must be developed. It is difficult to generate such interest in many new systems, let alone one that is 'different'.
- Component life is still somewhat unknown. Much of the technology used can be viewed as a 'black box' it either works or it doesn't. Factors beyond the control of the system may be responsible for some failures. The sensor units and controller boards in use have low failure rates so far; some have been in almost continuous usage for the last four years. Experience will show if component life is a problem in the long run.

Although these negative factors are small in number, they point out that this gauging system is merely a part of the whole manufacturing system. Much thought and preparation is required to bring any such unit on-line. For those willing to adapt, plan, and learn, this system will perform beyond their expectations.

THE EVOLUTION OF AUTOMATIC MONITORING AND DIAGNOSTICS IN AUTOMOTIVE MANUFACTURING SYSTEMS

V. Nicolo
Fiat Auto SpA, Italy

In any effort to improve the efficiency of operations, higher investment in plants allows greater scope for improvement. Monitoring and diagnostics are means of improving efficiency but they have to be conceived as an integral part of a manufacturing system. Fiat's early experience in this field and the recent achievements of the Termoli plant, where the engine FIRE is made, are described in this light.

The automotive industry, especially in the post-war period, has shown a very marked trend towards automation. The first developments in this direction, until the early 1970s, were towards highly automated machines that were capable of very high outputs of a single product. In fact, this was the only kind of equipment conceivable with the technology available, and yet this concept of automation was in accordance with the needs of the expanding markets at that time.

The automation was rigid in nature, but the rate of increase in market demand gave economical justification to the related investments. Return on investment was ensured by high volumes of output. In the mid and late 1970s and in the early 1980s, the Western world and, more specifically, Europe experienced a change in market attitudes. Customers became more and more sophisticated, demanding a greater variety of products of a high quality. At the same time, worldwide sales of cars sold were decreasing or, at most, were largely stable, and the forecasted market trends for the next decade look no more favourable.

Rigidly automated equipment could have been a hindrance to the managing of the new situation. Moreover, productivity was vulnerable to the widespread wild-cat strikes that were, in Italy in particular, marking a period of bitter labour relations in the late 1970s.

Fortunately the technological revolution, led by the development of microelectronics, makes it possible to conceive a change in the quality of factory automation, as follows:

- New, programmable automated equipment is available.
- More information processing capabilities exist in plants.
- Manufacturing lines and machine cells are given the capability of unattended operation for significant periods of time.

These developments render compatible objectives that previously appeared to be inconsistent. Such objectives include:

- Faster renewal of models against prolonging the exploitation of the investment made in equipment.
- Greater variety in the product range against higher quality levels.
- Quicker responsiveness to the market demand against reduction of stock in hand.
- Higher productivity against lower investment risk.

As a consequence, this new form of automation has been, in latter years, a clue to the strategy of some automotive companies. The degree of automation in car manufacturing has been increasing (Fig. 1),

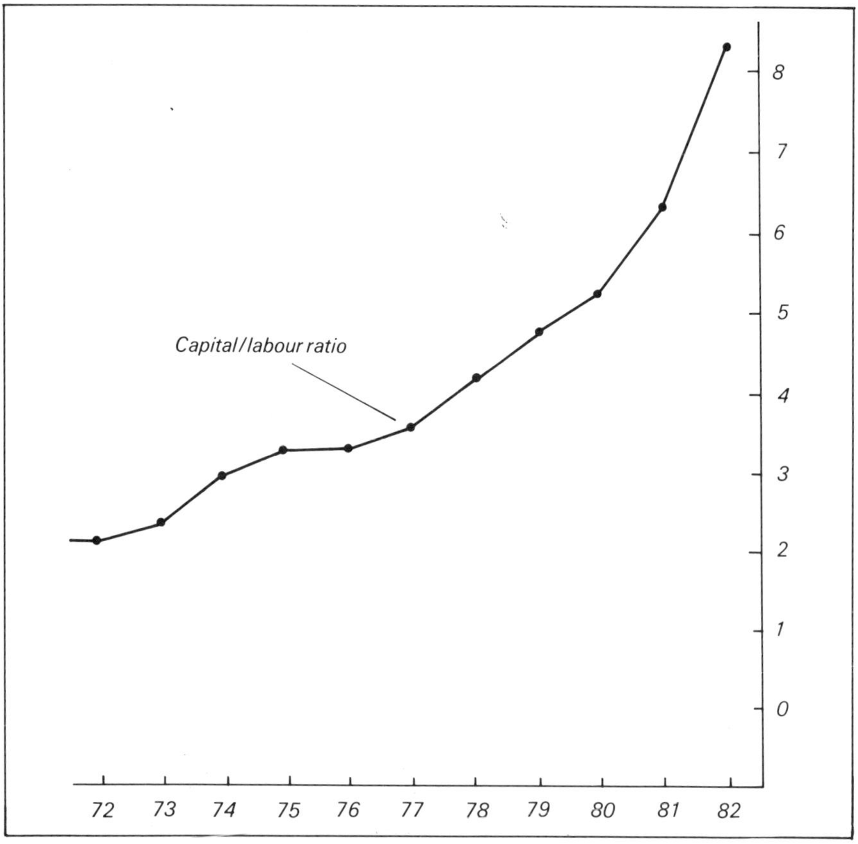

Fig. 1 Capital/labour ratio in Japan's automobile industry

although the situation is far from being uniform throughout the automotive industry.

The role of monitoring and diagnostics

The highly automated automotive manufacturing plants of today show a number of features and trends that increase the role of functions such as monitoring and diagnostics.

Relevant features and/or trends of the new automotive manufacturing plants are as follows:

- The investment value of the equipment is higher than in the past and will tend to become even higher in the future.
- Plants are more integrated and interconnected.
- The operating behaviour of the equipment tends to be more complex and the plant operations tend to become less deterministic.
- The workforce assigned to a plant is smaller than in the past and is tending to decrease (Fig. 2).

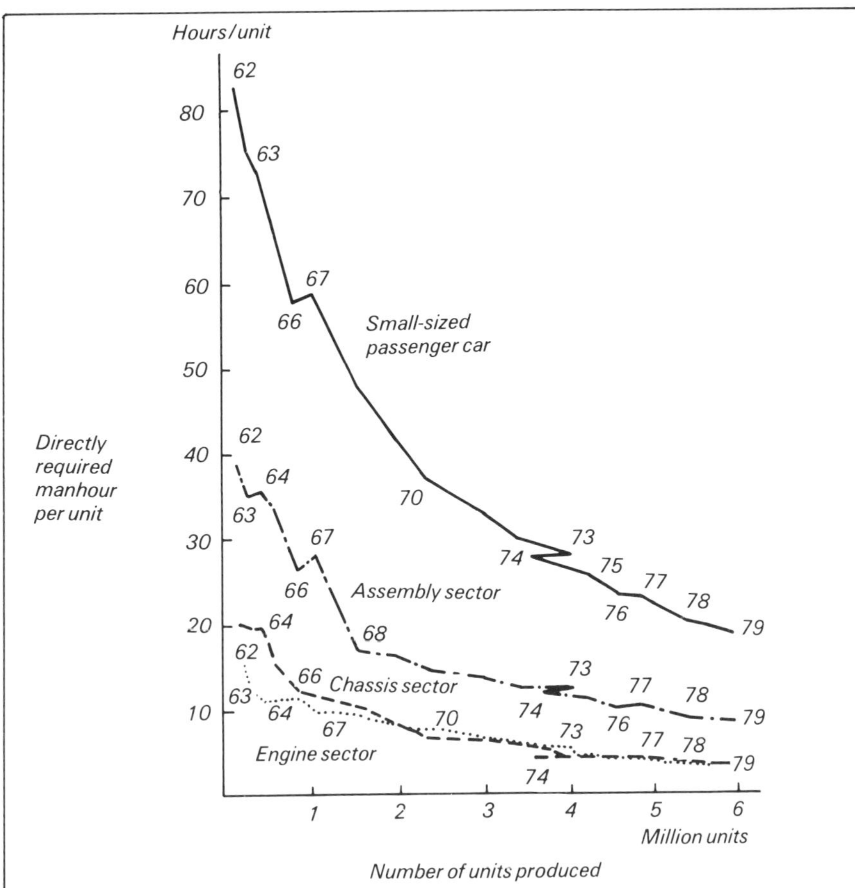

Fig. 2 Decrease in man-hour per unit of small-sized passenger cars in Japan

- The ratio of maintenance and supervision operators to direct workers is increasing.
- Electronics is in many instances taking the place of mechanics and, more generally, technologies which automotive workers were traditionally at ease with, are replaced by others which require training and a building-up of experience.
- Changes in the operating mode of manufacturing systems are often performed by typing commands on a console rather than acting directly on the machine command levers and knobs.

As a consequence of the above points:

- Plant and equipment down-times might produce greater losses.
- When accidents or abnormalities of the plant occur, the more complex situation might result in a longer time being required to discover possible causes and consequences and establish a correct resetting procedure.
- Further increase in overall man-hour productivity is more likely to be achieved by increasing the efficiency of those assigned to maintenance and supervision than by reducing direct labour through further automation.
- System faults and ill-functioning devices are often not simply detectable by visual inspection and workers cannot rely on past experience.
- Erroneous procedures do not lead directly to immediately verifiable and corrigible inconsistencies as, for example, mechanical interferences.

For all these reasons, it is apparent that monitoring systems, that keep the plant under observation, and diagnostic systems, that help maintenance to find failures, are necessary tools to improve the situation and to provide a smoother and more efficient and effective plant operation.

The more sophisticated the plant, the more sophisticated the monitoring and diagnostic system. If the plant is highly interconnected and integrated, it usually has a hierarchy of supervision levels, while its overall performance depends on a variety of interconnected functions. This implies that disturbances have far-reaching effects. In this case it is desirable that the monitoring system be capable of anticipating any possible hazardous situation that can be foreseen from the collected data but is possibly hidden by the complexity of the system.

On the other hand, warning and/or alarm flash lamps may be satisfactory monitoring tools for single machines embedded in a loosely connected environment.

In the same way, if the plant is highly automated, the data should be automatically and directly gathered from on-board sensors placed on the machines while only additional and/or statistical information should be left to operators on the floor for manual key-entry.

The final aim is not just to collect data but to run the factory as smoothly as possible, minimising the losses caused by any unpredictable disturbance. For this reason, monitoring and diagnostic systems tend to expand their original function - i.e. recording events and providing information on the causes of machine and/or systems failures - so as to include also guidance functions. These again may be developed as an aid to different levels of responsibility. At the lowest level this function provides simple operator guidance, for instance, recalling machine re-start procedures. This guidance is particularly important when a large number of machines are under the supervision of just a few people. At higher levels the guidance can be devised as an aid to help correct decision making, for example, on:

- Use of shopfloor storage.
- Allocation of available manpower at the beginning of each shift.
- Allocation and composition of maintenance crews.
- Programming tool changes.
- Rearranging production sequences to meet objectives.

Technical facilities

In order to support the functions that have been briefly described above, specific technical facilities are needed to handle the data input, the man-machine interface, the data storage and retrieval, the communication and the processing. In the last decade the advancement of technology in these areas, in the wake of the development of electronics, has been remarkable. Sensors offered on the market cover a very wide range of applications and have smaller dimensions and higher reliability than in the past. New technologies such as laser technology have made in-process measuring of characteristics possible, which previously could not be kept under continuous control (Fig. 3).

New low-cost technology has become available for displays and screens, which now offer graphic capabilities for a prompt visual communication with operators. Moreover, voice systems are undergoing rapid development.

In addition, computing power is now available at an acceptable cost to simplify the man-machine interface problems, providing operators with menus, display masks, support routines and the like, to the point that interacting with the system is no longer a specialist task.

Wired units are substituted by programmable logic controllers that sometimes have built-in diagnostic features. Low-cost, high-capacity, reliable memory devices, such as bubble memories and high-density floppy disks, are also available. Database and computer networking have become established technologies.

Nevertheless, the available technical facilities have to be arranged in suitable architectures so as to fit in with the needs of a specific situation whether it is monitoring a single stand-alone transfer machine, a transfer line or even an integrated plant. A number of problems arise

Fig. 3 Laser inspection of cylinder surfaces

from this. Some operating machines might not be pre-arranged to fit in the sensors capable of producing the required signals. Different equipment may have controllers that are quite different with regard to:

- The number and complexity of the functions that they handle.
- The programming language.
- The features of the operator console.
- The communication protocols.

Console units may have built-in monitoring and diagnostic features that have been developed at different levels making it difficult to standardise the functions to be given to a central monitoring system.

Here standardisation is particularly important. Nevertheless, users, besides making a general request for standardisation, are often obliged to comply with the available equipment and to make effective use of it. To this end, understanding their own real needs and the factual capabilities of the available technology is of paramount importance.

Fiat's early experiences

A preliminary experience with monitoring was made at the Fiat Mirafiori plant with an experimental system that was operating in late 1975.

The system, known as 'Tool Monitor', was developed with the cooperation of CSDL and was able to collect production and tool-cost data from individual machines. There was a central computer and mobile unit with no computation power. This unit, which had a keyboard and an alphanumeric display, was only meant to be an input/output device and to provide the interface to the machines. The signals automatically gathered from the machines concerned cycle-time, power on/off, piece presence and count.

Following this experience, a more advanced system was conceived as a means of obtaining a limited amount of production data automatically from an entire line of existing transfer machines. The system, known as 'Line Monitor', consisted of a PDP-11/34 computer without mass storage, several terminals and an interface to shop signals and was installed in Spring 1977 on the 131 Engine Block Line (Fig. 4). It produced a report for each shift on the performance of 27 machines and on the results of the final inspection operation at the end of the line.

The system was subsequently expanded: a second PDP-11/34 computer with disk and tape storage, a line printer, and additional terminals were added.

The original computer continued to collect data from the transfer line but it was now able to receive its programs from and transmit its collected data to the second computer. This expansion obviously added more features to the monitoring systems, such as:

- The ability to collect data and generate reports for periods of a day, week or month.

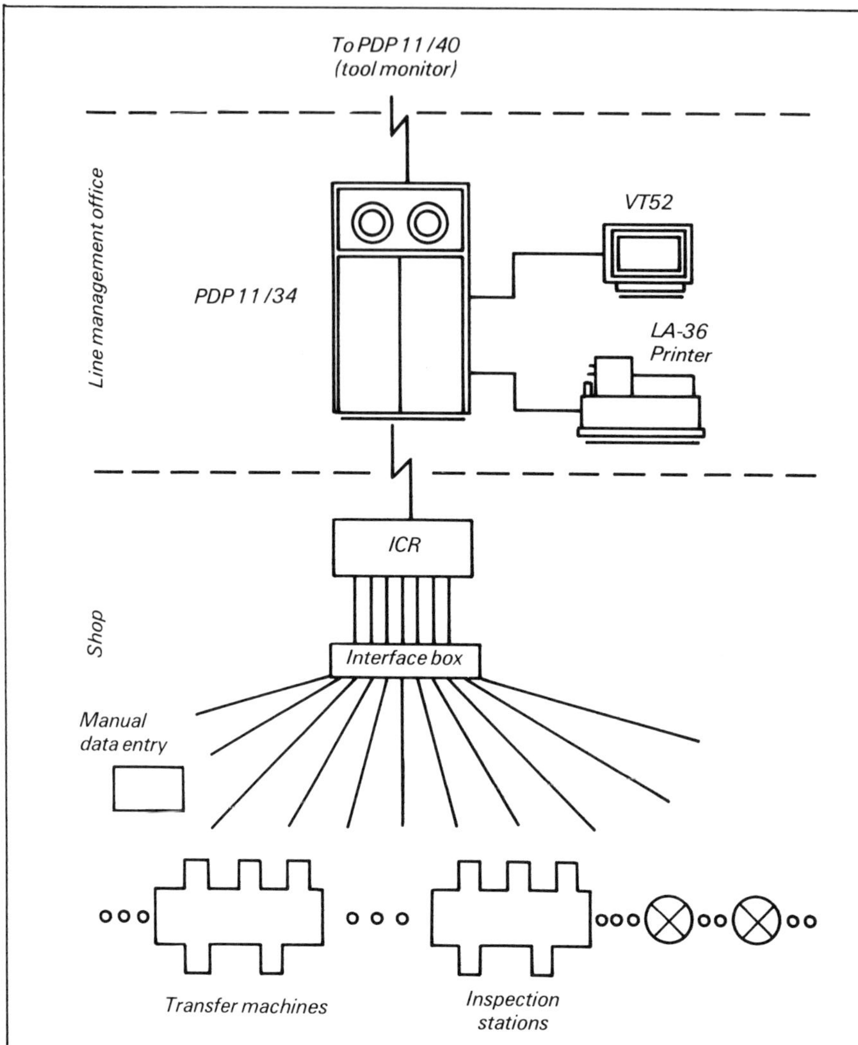

Fig. 4 Line monitor installation on 131 engine block line

- The ability to produce a large variety of more easily used reports.
- The ability to track and display the contents of buffers between machines.
- The capability to collect stop length distribution data.

In 1979, the two-level hierarchy was again adopted for a new monitoring system that was known as 'Multimonitor' and was developed with the support and the cooperation of Fiat Research Centre.

The scope of the system was to collect data relating to the operating cost of tooling on transfer lines, such as data on wear, breakage, tool change, sharpening and re-utilisation.

The data was collected by means of portable local computers, both automatically through the normal control signals of the machine and manually by operators. The operators were guided in the data entry operation by messages that appeared on alphanumeric video displays.

The local portable computers, protected by ruggedised cabinets and mounted on wheels, could be located remotely from the central computer on full-duplex telephone lines. This ensured the greatest portability of the local units that at the end of a data collecting campaign were disconnected and moved to the next line. Campaigns typically lasted three months and the number of lines connected at the same time was at most three. Other applications were carried out in the following years.

The experience gained from these pilot systems has shown that general-purpose monitoring applications require the acquisition of large amounts of very detailed data and as a consequence high sampling and communication data rates are involved. These requirements are specific when monitoring variables related to detailed motion rates, tool wear in-process detection, weld sensing and adaptive process control.

The pervasive access to this type of information cannot be accomplished just with add-on systems but requires the process to be controlled by computers or by control units that can easily interface them.

In other words, the monitoring function has to be conceived at the manufacturing plant conception time as an integrated feature of the total concept of automation. This has been the guideline in the new plant of Termoli 3.

The case of Termoli 3

Termoli 3 is the plant where Fiat makes its new concept small four cylinder petrol engine FIRE (named after the production processes used in its manufacture, i.e. 'fully integrated robotised engine'). The FIRE (Fig. 5) has just appeared on its first utilisation on the new Autobianchi Y10. This new engine, that has important innovative features with respect to its immediate predecessors of the same capacity, has been specifically designed in such a way that it can be economically produced with highly automated production processes in all the technological areas involved, including engine assembly.

The plant, located near Termoli in Southern Italy, is designed to produce not just one but several different versions of the engine. In the same plant all units are assembled and all main components machined: cylinder blocks, cylinder heads, and associated intake and exhaust manifolds, drive shafts and connecting rods, camshafts and timing system components.

The automation level reached is very high also in the engine assembly: virtually 100% in the cylinder head and in the connecting rod/piston assembly areas, 80% in the overall assembly. Quality control is given the

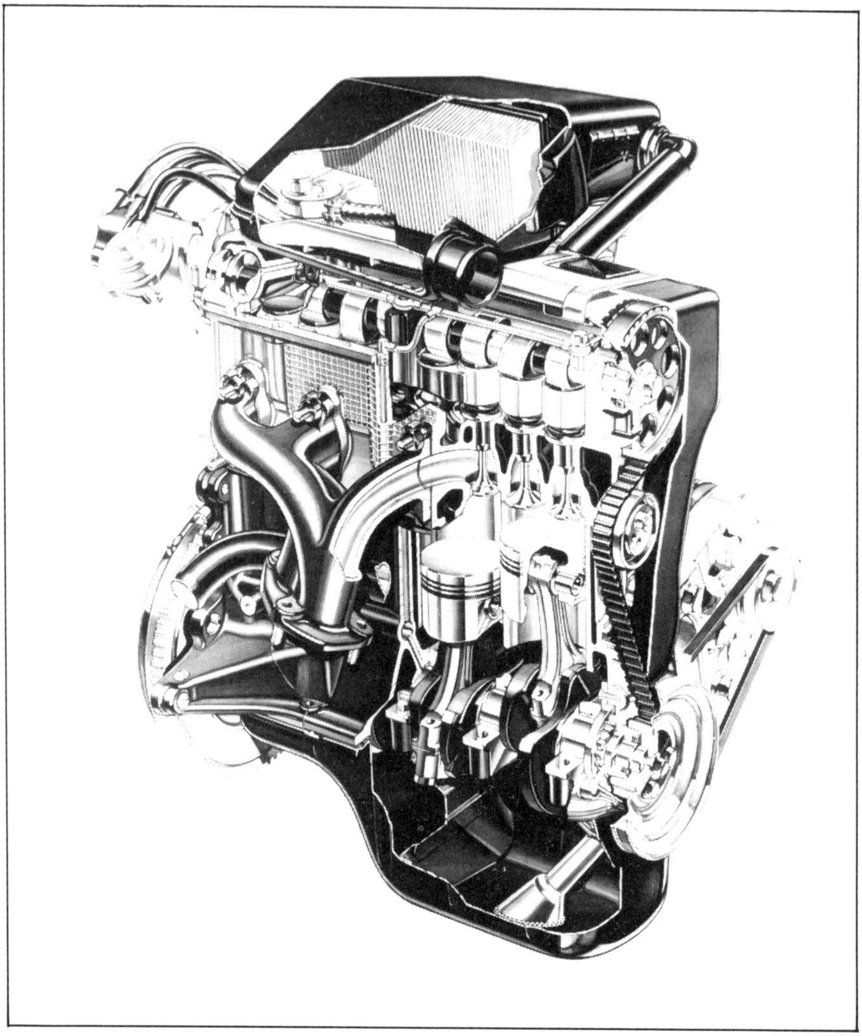

Fig. 5 The engine 'FIRE'

greatest attention. 'In-process' control has been used whenever possible and the engines are put through a completely automatic test cycle (Fig. 6).

Material handling is automated throughout the machining area. Automatic systems for recognition of components are widely used to keep track of the flow in the assembly area. The investment required for the plant has been around $300 million.

Owing to the high level of automation and interconnection, breakdowns or operating instability could cause considerable damage. Here, monitoring and diagnostics are no longer just an added feature of the plant but have become an integral part of the total concept of automation which in Termoli has been implemented, breaking new grounds:

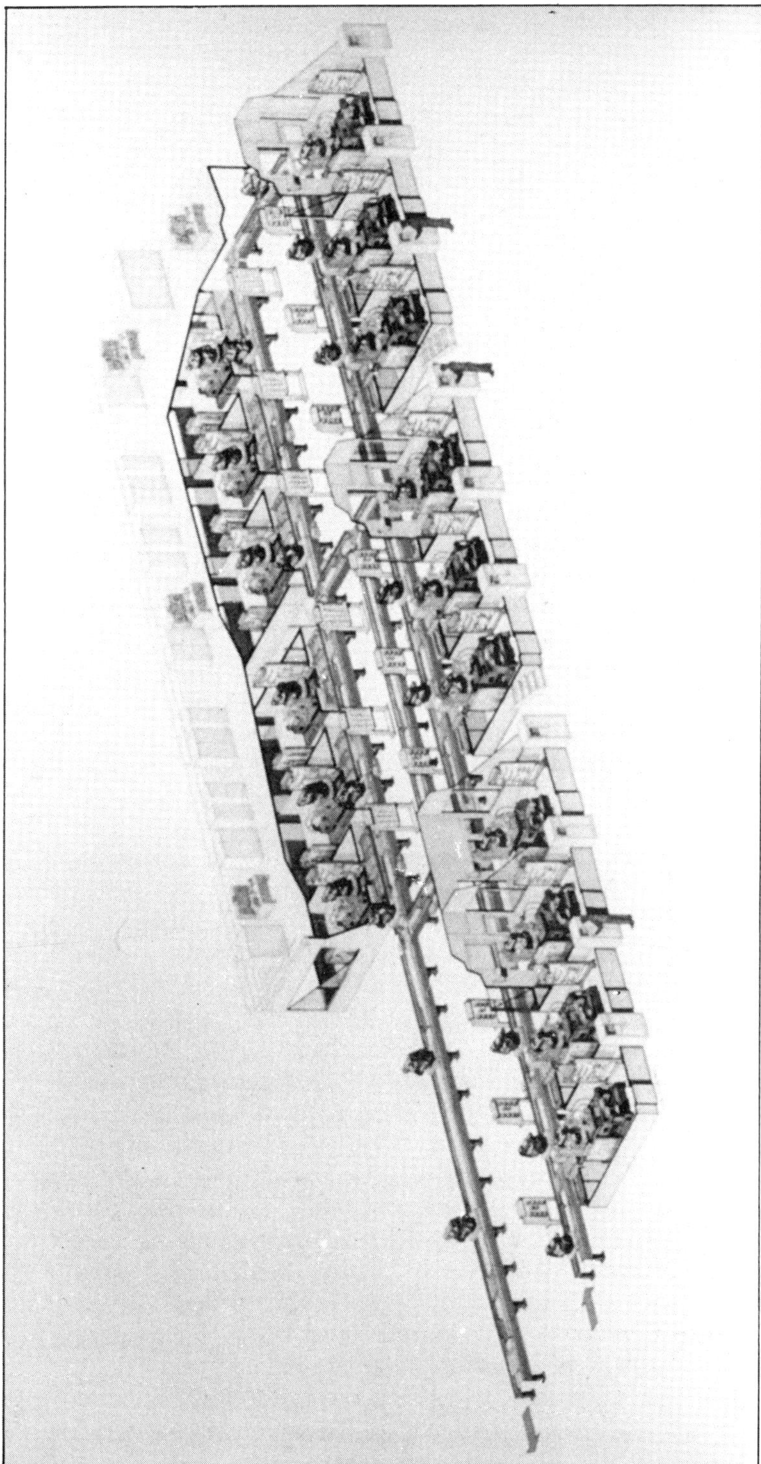

Fig. 6 Termoli 3 layout of the engine testing area

- The processing equipment in transfer lines is all servomotor driven.
- Devices are installed to safeguard the quality of critical tolerances so as automatically to adjust those machined dimensions that are particularly subject to change as a result of tool wear, by using in-process and post-process measurement signals.
- The other dimensions, less subject to variation, are checked by means of computer controlled on-line automatic statistical control, by using multi-purpose robot controlled measurement devices.
- Automatic tool replacement is adopted when frequency and heaviness of duty make the feature significant in terms of reducing down-times.

To reduce down-times further, machines are connected to computerised monitoring and automatic diagnostic systems. These have been organised on a two-level hierarchy: a local area level and a central level.

The equipment included in the local area does not only refer to the main operating machine but comprises also ancillary equipment around it, such as the part transfer system from and to the machine, the machine loading/unloading system, the part inspection and testing unit, the tool-wear verification unit.

Local monitoring and diagnostics are always associated with local flash lamps signals to catch the attention of operators on the floor. These indicate three possible conditions of the machine: automatic operation, alarm, and warning. The alarm condition is not necessarily due to a machine fault but might also be generated by the lack of parts of any other machine stop cause not related to the incorrect functioning of the machine.

The causes of alarm and warning state signals are reported by the local diagnostic system which has a different level of sophistication depending on the complexity of the associated machine. The diagnosis can get as far as indicating the single station and/or unit responsible for the warning or the alarm state and the phase of the machine cycle involved. The functions monitored by the local area monitoring systems include: operator instruction systems, display of functional cyclograms, tool change monitoring, programmed maintenance monitoring, self-diagnosis, statistical failure analysis and statistical productivity parameters.

Local monitoring is carried out by means of a variety of systems, although standardisation has been pursued as much as possible. In fact, as previously pointed out, local level monitoring and control functions are very close. On the other hand, some machines or devices come in a single package with their control units that in turn may have built-in diagnostic and monitoring features. Nevertheless, the tendency has been to standardise at a performance level such as that featured by the Comau Hermes system that has been widely used throughout the plant, the capabilities of other available systems being increased by the central monitoring system.

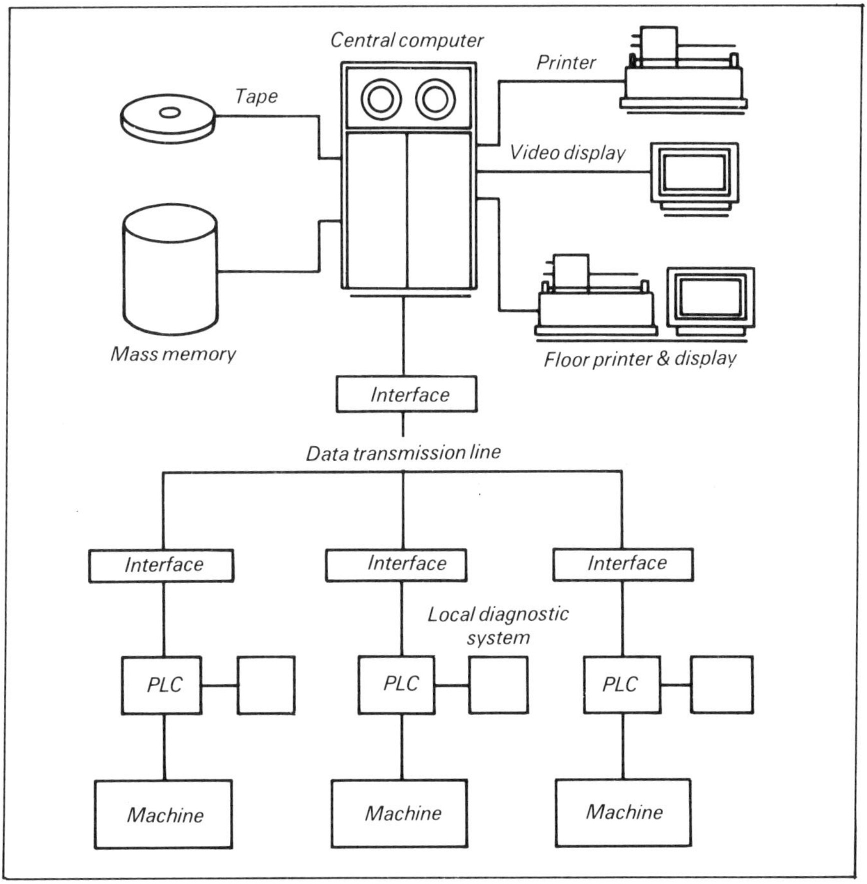

Fig. 7 Central monitoring system

- State of the system (2)
- Machine productivity (1)
- Machine stops (6)
- Line efficiency (4)
- Cyclogram related data (1)
- Buffer storage levels (1)
- Planned maintenance (3)
- Shift classification (1)
- Planning reports (7)
- System messages (1)

Numbers in brackets refer to the number of different kind of reports and/or displays available.

Fig. 8 Central monitoring system displays and reports

The setting up of a centralised monitoring system (Fig. 7) is currently in process. The central system is designed to give direct access to suitable and up-to-date reports to the people who are in charge of manufacturing, maintenance, quality and plant services, with the aim of controlling the production flow (Fig. 8).

Concluding remarks

Some of the relevant points mentioned in this paper are summarised below:

- Higher investment in plants gives more scope in any effort towards more efficient operations.
- Monitoring and diagnostic systems are means of reaching more efficient operations but they have to be conceived as an integral part of the manufacturing system.
- Local monitoring is similar to controlling in many aspects but control units often depend on the choices of machine builders. Standardisation is required to have well-balanced monitoring systems.
- The final aim being smoother plant operation, the expansion of monitoring systems with guidance features is a desirable trend.

INDUSTRIAL USE OF ULTRASONIC RANGING SENSORS IN ROBOTICS

M. Guichard
Peugeot SA Cera, France
and
A. Renault
AKR, France

Extensive use of ultrasonic ranging sensors in industry has been made possible by today's available technology. The characteristics of these sensors such as accuracy, non-contact, size and low cost make it possible to study their possibilities in the field of robotics, both experimental and practical applications in industry. An application of car body relocation to suit the path of a spraying robot is described; others are also discussed.

New sensors and computer technology are now available for industrial application. Not surprisingly, mechanical equipment is progressively being replaced by technology which sometimes seems more sophisticated to people not used to it. These new solutions to manufacturing problems offer more flexibility and sometimes at a lower cost.

Performing sensors are being introduced into the automobile industry to help operations on the line. Some need to be tested in this new environment. Ultrasonic sensors already belong to the industry, and they can offer a wider range of applications, not only in detecting presence, but for measurement and robotic systems.

Ultrasonic ranging sensors have been chosen to be part of such an application at Peugeot SA. In this situation a car body must be located in order to adapt in real-time the path of a spraying robot. They are now in use in some plants within the automobile industry.

Principle of ultrasonic ranging sensors

An ultrasonic ranging sensor consists of a transmitting and a receiving acoustical transducer both of which might be contained within the same unit.

The transmitting transducer sends a pulse towards a target and a resulting echo is detected by the receiver (Fig. 1). The elapsed time between the initial transmission and the echo detection is proportional to the distance to the target. The pulse is emitted within a beam angle and only the first echo emerging from the nearest target within the beam is detected.

To measure distance using ultrasonic ranging sensors, the speed of wave propagation in the air is known as:

$$V = Vo\sqrt{1 + \Theta/273}$$

where Vo is speed at 0°C and Θ is temperature in °C. This is because measurement of distances with ultrasonic sensors is temperature dependant.

Application information

Depending on the type of sensor and its manufacturer, ultrasonic ranging sensors can measure from a few millimetres to metres, with a level precision beginning at 0.1mm.

For robotic applications, this type of non-contact sensor is compact, light, rather sturdy and immune to noise. They are available also at reasonable prices (from about $10) but require some additional electronic devices with which to interface.

However, they should not be placed in areas subject to air turbulence, for example close to air pressure outlets.

Robot path adaptation

Current manufacturing lines handling car bodies cannot provide the required positioning accuracy for some specific operations. This problem can be solved by one of a number of different ways:

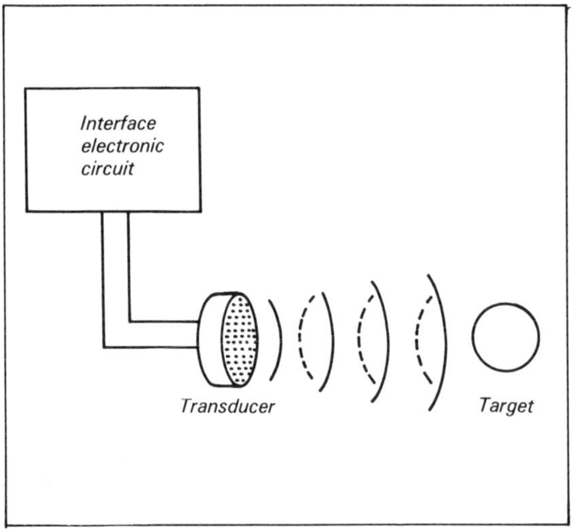

Fig. 1 Principle of ultrasonic ranging sensor

- By modifying the plant conveyor to achieve the required location of the car body at the automatic station to accomplish the task.
- Using a system of sensors to help the automation, and in particular the robot, adapt to the real position of the car body.

The second option has been chosen for an AKR robot applying sealant at the Peugeot plant in Mulhouse. Here there was a problem because once the robot had applied the sealant, surplus material on the doors then had to be removed. Also, occasionally some additional sealant had to be manually applied because of the variation in the location of the car body on the line. To modify the conveyor to cope with this seemed to be both expensive and not very flexible.

The requirements were as follows:

- Offsets between any point of the car body and an average position should be less than ±50mm.
- The final precision on the location of any point should be ±5mm.
- The variation in dimensions between successive car bodies is not to be taken into consideration.
- During the operation only the car body is moving along the conveyor (the swinging movement is neglected).
- The measurement system and the working location can be separated. The position of the car body relative to the conveyor does not change between the two locations.
- It is not possible to place anything on the car body or to measure it by making contact.
- The type of car body is known.

Ultrasonic ranging sensors with a 1mm precision and 30°C beam angle were chosen for the measurement system. Other alternatives included vision, stereovision, laser sensors and other types of ranging sensor. At that time, this type of ultrasonic sensors was easy to use and interface; it was also cheap compared to the other types mentioned.

Car body relocation

The location of any solid in space is known from six parameters: three for the position of the origin of the reference frame of the solid; and three for the orientation.

These six parameters can be inferred from the measurement of distances between six planes of the solid and six known fixed points[1,2].

In general the solution of the system is rather complex. Some simplification can be made, such as choosing perpendicular planes. It is also possible to set free from the variation of position along the conveyor by using a sensor to synchronise both stations (measurement and robot).

Therefore, five reference zones must be chosen on the car body such that:

- The zones should be as large as possible and at least equal to the variation in car body position.
- The distances between the zones should be as large as possible (for improved precision in angle computation).
- Planar zones should exist within perpendicular planes.

The sensors must be set such as to cater for the medium position of the car body:

- The aiming axis of the sensor should be perpendicular to the reference zones.
- The sensor-to-plane distance should be as small as possible.

However, the smaller the zone, the further must the sensor be to stay within the aiming beam (see Fig. 2). Fig. 3 shows the position of the sensor system when the car body is in the measurement location.

To modify the robot path, the position of every element of the operation must be known to a high degree of precision. The robot trajectory must be considered as part of the body frame. Therefore the trajectory must be varied by the same transformation as measured by the sensor system.

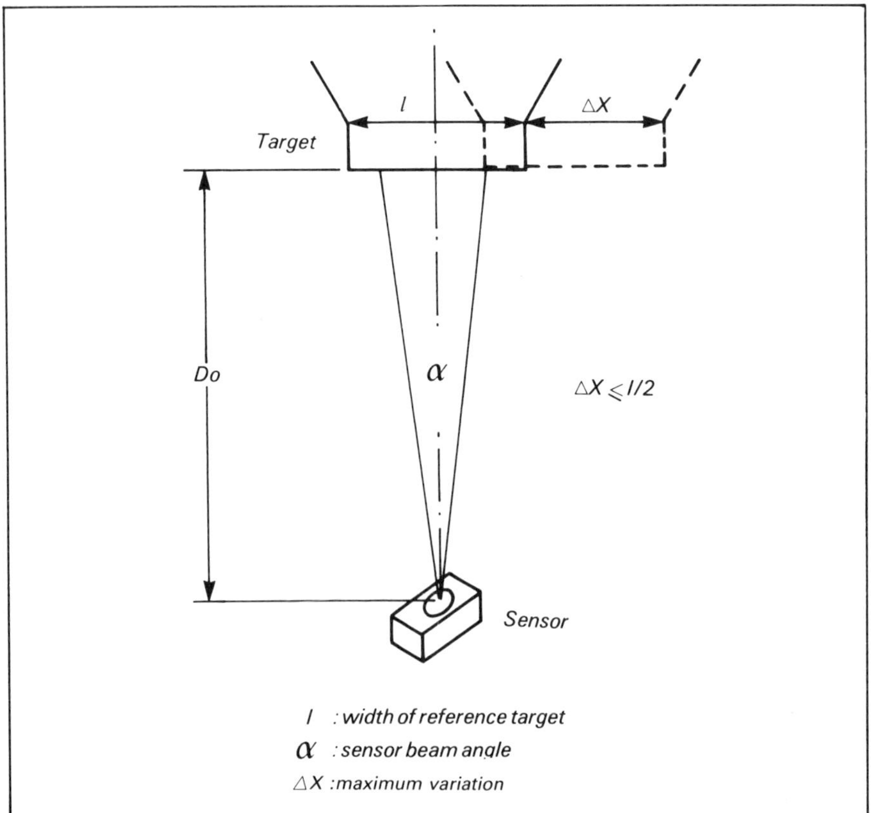

Fig. 2 Aiming beam and target location

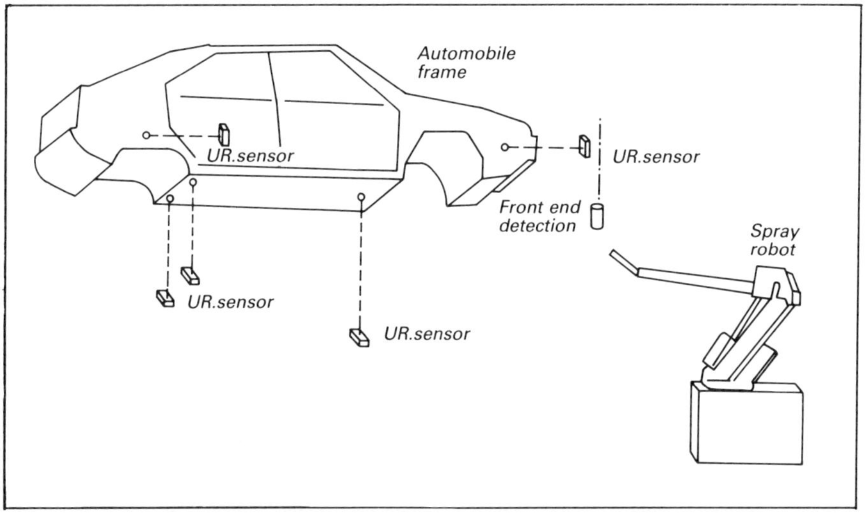

Fig. 3 Location of ultrasonic ranging sensor system

The conveyor frame is taken as reference frame. The robot is moving along with the conveyor, and its frame must be defined into the conveyor frame. This is a common method of synchronising robots to achieve a good precision range.

Expressing the sensor system frame into the conveyor frame is a critical operation because the sensor emitting and receiving centres are not precisely located.

To lower the influence of these values, it is proposed to measure the offsets from a reference.

Operation and measurement

The experimental study of the system was conducted by Peugeot SA within its CERA (Centre d'Experimentation en Robotique et Automatismes) in cooperation with AKR, the robot company. AKR developed the specific software for the controller of the robot 3000 and for the interface with the measurement station.

A programmable controller is used to interface the sensors with the robot controller and monitor the following steps of the operation:

- Detection of the front end of the car body using a sensor, although this does not need to be ultrasonic.
- Monitoring and checking of acquisition cycle by ultrasonic ranging sensor system.
- Acquisition of the temperature information for the correction of the measured distance values.
- Transmission of data to the robot controller through an RS232 port.

The robot controller, providing the five distance parameters, computes the offsets using the car body reference distances. Then some preliminary data are computed and the robot can start its modified trajectory. Each point is altered in real-time with the offsets on *X, Y, Z*, yaw, pitch and roll.

The sensor system has to be located in front of the robot in order to allow time to send the five distance parameters (at 9600 baud rate) to the robot controller and for some preliminary computation (total about 1–2s). The sensor system has to be some distance away to avoid clogging by the sealant drops.

Several experiments have been carried out showing satisfactory offset correction. The maximum error in the location of the point of the robot on the car body is within the required 15mm.

A computer simulation was implemented during the same period of time. It could not take into account real factors such as beam angle, offsets interactions, error in conveyor position, and so on. On the other hand, a sensor with a precision of ±1mm was simulated rather than the more precise sensor (1mm) used in the experimentation. This simulation showed a better correction (error within ±3.5mm). So it was concluded without experiment that there was every possibility this method can compensate for a ±50mm variation in car body location within a error range ±5mm.

This computer simulation is also showing that the sensor location has little influence and that more accurate sensors can greatly improve the error correction.

Concluding remarks

Varying the path of a robot with the help of a sensor system has been achieved within the manufacturing specifications. The robotic solution is flexible and the overall cost several times lower than for mechanical equipment (sensors or sets of sensors can measure several car body types).

A number of successful experiments with ultrasonic ranging sensors suggests they may have a wide use in the robotic field, such as:

- Centering of inside roof cover. Two sensors on the end tool of the robot are measuring in differential the interior side of the car body. When the difference reaches zero, the cover is centered and can be pressed.
- Location of a pallet.
- Location and orientation of parts in a container to be unloaded.

References

[1] Andre, G. 1983. Conception et modelisation de systemes de perception proximetrique. Application a la commande en teleoperation. Thesis, Docteur Ingenieur, Rennes, October 1983.

[2] Albus, J.S., Barbera, A.J., Fitzgerald, M.L., Nagel, R.N. and Vanderburg, G.J. 1980. A measurement and control model for adaptive robots. In, *Proc. 10th Int. Symp. on Industrial Robots*, pp. 303-312. IFS (Publications) Ltd, Bedford, UK.

[3] Karoyt, T.B. 1985. Conception et realisation d'un proximetre et telemetre pour utilisation en robotique. Thesis, Docteur Igenieur, ENSAEE, Toulouse.
[4] Monchaux, S. and Prat, R. 1984. Telemetrie acousto-optique pour robot mobile. *Le Nouvelle Automatisme*, June 1984.
[5] Kanade, T. and Sommer, T.M. 1983. An optical proximity sensor for measuring surface position and orientation for robot manipulation. In, *Proc. 3rd Int. Conf. on Robot Vision and Sensory Controls*, pp.301-308. IFS (Publications) Ltd, Bedford, UK.
[6] Paul, R.P. 1981. Robot Manipulators Mathematics Programming and Control. MIT Press, Cambridge, MA, USA.
[7] Biber, S., Ellin, S., Schen K. E. and Stempeck, J. 1980. The polaroid ultrasonic ranging system. In, *67th Convention of the Audio Engineering Society*, New York, USA.

5

FLEXIBLE MANUFACTURE

Flexible manufacturing cells have become increasingly popular in the automotive components industry as vendors strive to meet demands for greater product variety, shorter production runs, faster set-up times and shorter changeover periods. A number of companies have done much to advance the technology and establish worthwhile experience which has proved to be of benefit to others. However, systems are by no means perfect and there is still much to learn, with the financial justification of FMS being one with which many have still to come to terms.

THE EVOLUTION OF FMS AT VOLVO

G. Nordsten
Volvo Components Corp., Sweden

FMS increases productivity, capital utilisation and reduces capital tied up in inventory. At the same time it increases the possibility for adjusting the production to follow fluctuations in market demand. Volvo Components Corp., Engine Division in Skövde, has installed a large number of FMS installations during the last five years. Four of the most important installations in the company are discussed here. The FMS installations for flywheel had to be rebuilt because of insufficient capacity for the future. The differences between the old and the new installation and the evolution into higher flexibility and productivity in the new installation is described, as well as the way of involving the operators in the project.

Volvo Components, Engine Division in Skovde, Sweden is a part of the Volvo Group. The company was established in 1868 and was at that time a producer of castings and machine tools.

In 1907 the first combustion engine was produced. The first engine for the Volvo car was produced in 1927 - a four cylinder engine with an output of 28bhp at 2000rpm. The expansion and technical development since that time has been tremendous. Today Volvo Components Corp. manufactures and supplies the drive-line to Volvo. The company consists of three divisions: Engine, Transmission and Blanks Division. The Engine Division has roughly 5000 employees and an annual turnover of more than SEK 400 million. The main production is more than 300,000 petrol engines and 60,000 diesel engines a year.

It is considered to be one of the leaders in using new technology but that alone is not sufficient because sooner or later competitors will catch up with the technology. Therefore Volvo considers that its most important asset is the employees. Together with them and the unions Volvo has agreed upon an upgrading of everyone's competence and commitment. Volvo's goal is to be the leader in quality, flexibility, productivity and delivery reliability.

Automation and economy

The engine has achieved a productivity increase of more than 8% per year over the last nine years. The investments have reached SEK 500 million in 1984. Volvo Components is today probably one of the investment giants of Sweden. It was estimated that in 1984 Volvo Components Corp. accounted for 4% of the country's total industrial investments. Quite a lot of the invested money has been used for automation in so called flexible manufacturing systems. The main purpose with this type of automation is to improve capital turnover through high machine utilisation and short lead-time. In Fig. 1 the difference between production with machine tools grouped functionally and workpiece orientated production with FMS is quite obvious.

Other advantages include:

- Simplified planning.
- Shorter and more secure delivery time.
- Better space requirements for stock.
- A better overview of both capacity and disturbances.
- Earlier discovery of faulty workpieces.

The investment cost for FMS can be high because of the demands of high utilisation which among other things means automated material handling, checking and monitoring devices. If business declines, or there are major changes of products, problems would arise because these FMS are often less flexible in the handling systems than manually loaded conventional machine tools. Therefore, it is important to obtain

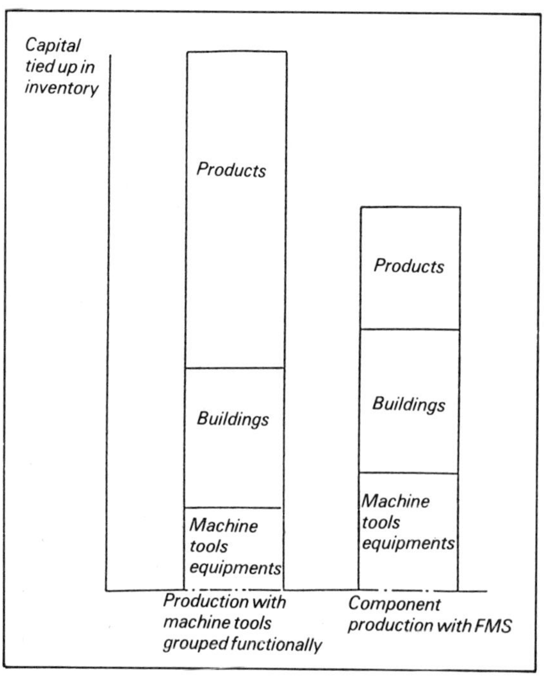

Fig. 1 A capital rationalisation is reached through FMS installations

equipment at a reasonable price and it should limit flexibility as little as possible. Volvo has a design department, both for mechanical and electrical designs; and this makes it easier to reach this goal than if all the equipment was to be purchased.

Designing the layout and training

To create a successful FMS installation it is necessary to involve more technicians from the start and to plan for a longer run-in time than when making an installation with conventional machine tools. As Volvo Components Corp. during recent years has carried through several FMS installations of many different types it now has extensive knowledge throughout the whole organisation. When designing a new FMS installation time schedules should not be too tight since delays and difficulties are quite common when running-in the systems and may, when it is a complicated installation, cause interruptions of as much as six to twelve months.

The basis for a successful FMS installation is created during the design of the layout. Besides analysing the types of machine tools; handling equipment, materials flow, control of individual machines and groups, communications nets, program storage, measuring equipment and alarm systems operating crews and need for training have to be thoroughly assessed. It is important to keep the technical solutions as simple as possible. Complicated and very advanced solutions can cause disturbances and reduce reliability. Education and training of the operators must start early to give them a sound knowledge of the installation and thereby help reduce run-in time.

Earlier installations

Since 1980 15 different FMS installations have been undertaken in Skovde (if an FMS installation is defined as at least two NC controlled machine tools connected to each other through an automatic handling system and controlled by a main computer). Most of the equipment has been produced within the Volvo Components Corp. This includes: multi-head drilling machines, NC crankshaft grinding machines and NC gantry loaders. The installations have been of different sizes and grades of automation.

Two different FMS installations have been made for manufacturing diesel engine crankshafts. The first one, for 10 and 12 litre engines, was installed in 1982 (Fig. 2) and the second one, for 6 and 7 litre engines, in 1984. The installations today for heavy crankshafts consist of seven different cells, each one with two machines and an NC gantry loader. The gantry loader has two separate grippers with a loading capacity of 250kg each. The gantry loads and unloads the machines with components brought in on pallets transported by an AGV (automated guided vehicle). There is always a two-pallet buffer stock between the

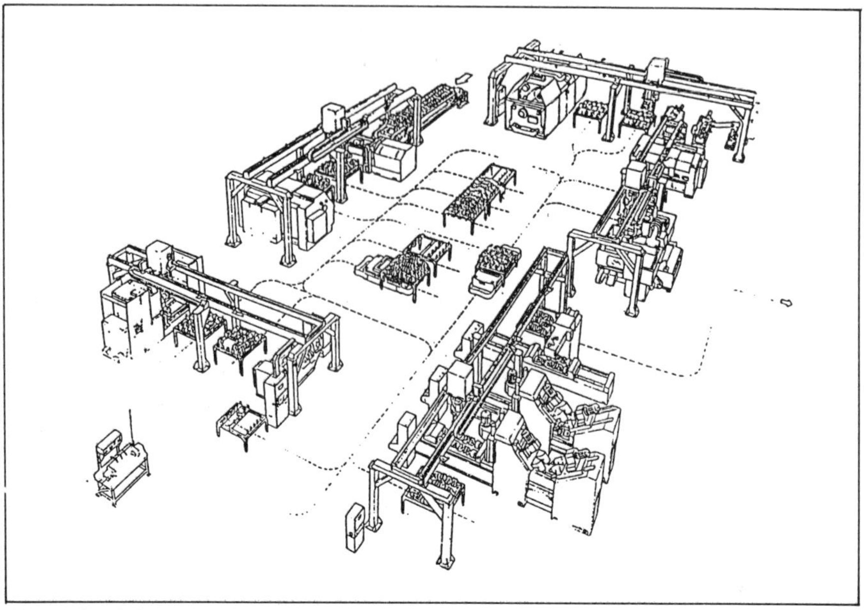

Fig. 2 FMS for the manufacture of diesel engine crankshafts

machines in order not to run out of components. Machines and handling devices are all connected to a main computer.

When designing these systems Volvo has, as far as possible, retained existing machine tools and therefore has been able to reduce investments. Through a flexible buffering place in the centre and by keeping solutions as technically simple as possible, Volvo has achieved more than 80% reliability. In the 10 and 12 litre engine crankshaft installations 30,000 components can be machined a year. For the smaller crankshafts the production rate is 23,000 a year. In both systems two different sizes of crankshaft can be produced in batches adapted to the need in the assembly hall.

Another installation that went into production in 1982 is an FMS installation for machining intake manifolds for the petrol engine. Design, manufacture and installation took 14 months. Four different groups, each served by an ASEA IRb60 robot, are installed on both sides of an accumulation roller conveyor for transportation of workpieces placed on a small steel pallet. Figs. 3 and 4 show the layout and technical specifications.

FMS installation for producing flywheels for petrol engines

Using the experience gained from, among others, the installations described above, an FMS installation to produce five different types of flywheel for the petrol engine was designed in 1982. The capacity was calculated to 20,000 flywheels a year from production in three shifts. It was installed during Spring 1983 and put into production in June 1983.

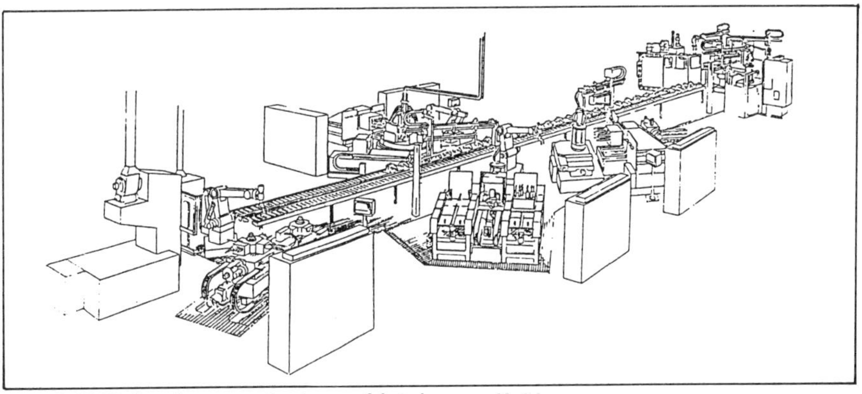

Fig. 3 FMS for the manufacture of intake manifolds

The flywheels are cast in Skovde and delivered directly from the foundry to the FMS group. Pallets containing the castings are transferred from the truck onto an in-transport conveyor and from there the castings are fed one by one into a magazine at the first Pittler lathe. The installation consists of three different groups each served by an ASEA IRb60 robot. For transportation between the groups, roller conveyors are installed on which workpieces for half an hour's

	New line	Old line
Retooling time between different manifolds	Half an hour/machine	20 hours/machine
Labour requirements	2 persons/shift	5 persons/shift
Unmanned production	At least half an hour	Not possible
Flexibility	8 different manifolds	Non existent
Capacity	250,000 manifolds/year	250,000 manifolds/year

Volvo flexible multihead machines. Drilling

Drilling head, width × height	780 × 700mm
Adaptor plates	780 × 720mm
Feed travel	630mm
Power at 1500r/min on output shafts	24kW
Feed force	60,000 N
Feed speed	10m/min
Shifting time for drilling and tapping heads	8sec
Control system	3-axis CNC

The machine can also be specified with larger or smaller dimensions

Fig. 4 Flexible manufacture of intake manifolds for petrol engines

production can be stored. The first group consists of two Pittler lathes for rough turning and a heater for the flywheel gear teeth.

The second robot group consists of a fine turning lathe, a camera for checking the surface finish and a multi-head machine for complete machining of the attachment holes. From this group, containing only new equipment, the flywheels are put on a powered accumulating roller conveyor which leads to the third group. On its way the flywheel passes a drilling machine (for indicator holes for a Renix ignition system) and a washing machine.

After these two operations the flywheel reaches a three-station gantry loaded assembly machine for inserting the pins in the holes for the ignition systems. There is also an automatic control of the heights of the pins after the insertion operations. The robot then takes the flywheel and brings it to a balancing machine and a manually operated checking station. If the quality is acceptable the robot places the wheels into a pallet in several layers with masonite boards between. The whole FMS installation is operated by three men per shift.

When starting up the system there were several small problems from the two lathes because of chips and tool breakdowns. The updating of the machines and the whole installation of the system also had to take place during a very short period. It was required to reach full capacity within three months of the start of production.

The layout is tight because there was no more space available; in addition, the Renix drilling machine was first decided upon when the installation was about to take place. The tight layout caused problems as it meant that there was no space for transporting material in and out of the system, given that a stoppage took more than 30 minutes to solve. The buffering capacity on the roller conveyors is limited to 30 minutes' production.

Since 1982, when this FMS was planned, the engine production has increased rapidly. Volvo has now moved the installation to increase capacity further and to reduce stoppage time caused by the initial system layout.

During January–March 1984 different proposals were made about how to solve the lack of capacity as well as problems with the old installation. Volvo chose to make the same arrangements as for the FMS installations for crankshafts, that is to say, separate cells with machine tools loaded by gantry loaders and connected with each other by AGV. This meant the system could be fed manually if anything should happen to the AGV system or gantry loaders. Also, if any one of the machine tools broke down components could be stored in the centre buffer.

The layout is shown in Fig. 5. In this new installation nine NC gantry loaders made by Volvo Components Corp., Alton Engine Division, are included. Most of the equipment from the old installation is used in the new one with the exception of the lathes which are replaced by new vertical machines from Diedesheim and equipped with standardised tooling. That means when changing from one type of flywheel to

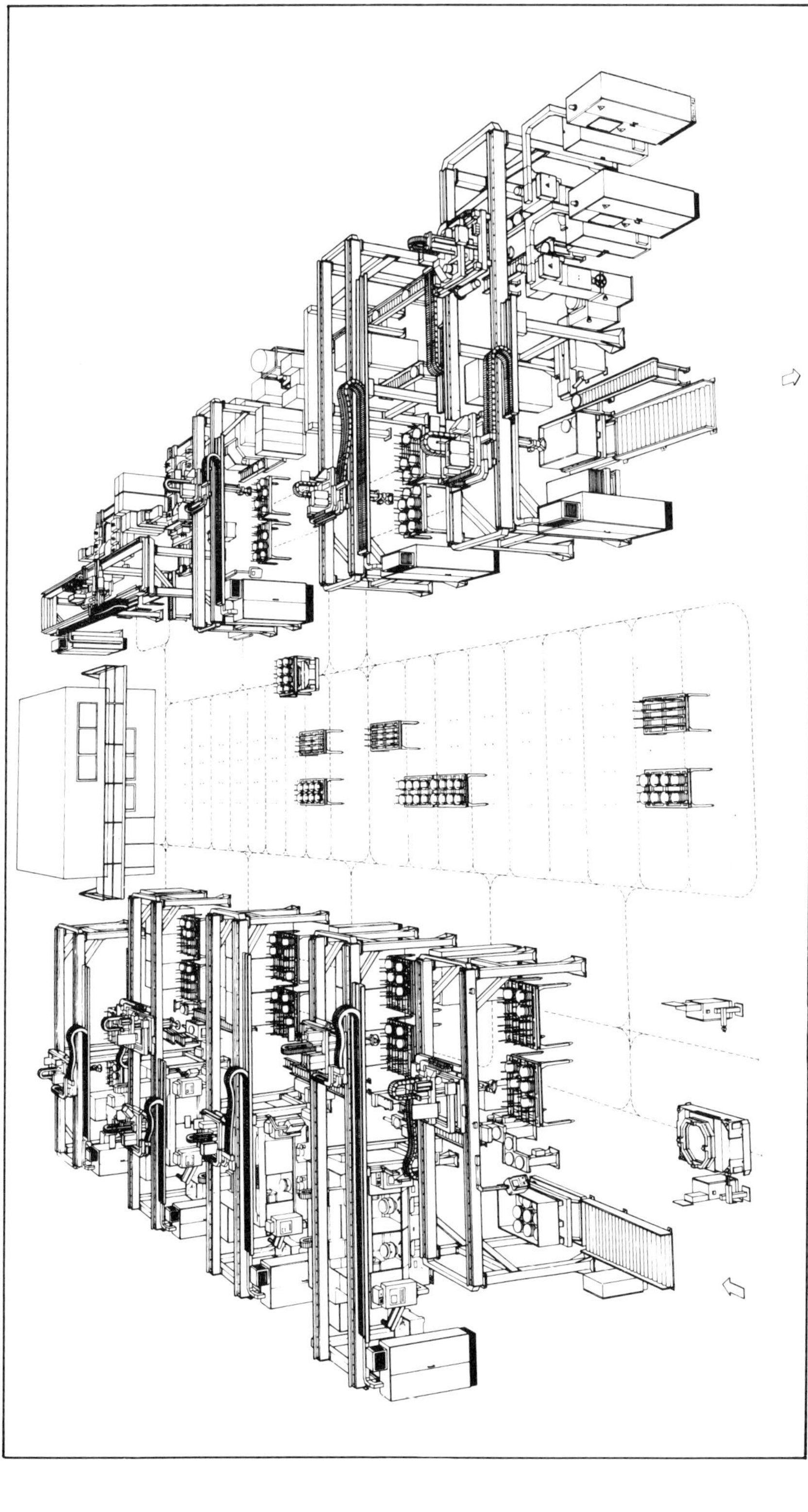

Fig. 5 Revised FMS layout for the manufacture of flywheels

another only the NC program has to be changed, and this can be done from the computer keyboard. Compared with the old installation two cells have doubled capacity through the introduction of a new multi-head drilling machine and a balancing machine.

Since AGVs are used for the transportation of pallets between the cells, the machines can be placed out of operational sequence but with due consideration given to the handling of chips and to the machine size.

The gantry loaders serve only one machine each except for the multi-head drilling machines and the balancing machines which are placed together. There are also checking devices and tilting devices which the gantry has to load. The machining time for one flywheel is a maximum of 47s, in each station.

Forty pallets are used for transporting flywheels within the system. On each pallet 42 flywheels can be loaded which are placed six on top of each other in seven positions - there is always one eighth position which is free. This means that it is possible to buffer wheels for at least 8 hours' production if anything should happen in a workcell. For transportation between the different cells two AGVs are used, with a third as a spare.

Checking after each operation is carried out through random samples; the checking frequency can be programmed. Pallets can be ordered out of the system whenever needed using the keyboard. Workpieces requiring re-adjustment can be brought back into the system also on pallets and then the transaction has to be ordered by the keyboard. Only workpieces that have passed the same operation can be placed in the same pallet.

The computer controlling the gantries and the computer controlling the AGVs exchange information when bringing pallets into and out of a cell automatically. Any change in the number of workpieces is then reported; all programs can be changed automatically. There are few tools to be changed so that retooling of the whole system takes a maximum of 20 minutes. Every unit is also equipped with an alarm system connected to the control system, reducing maintenance time.

Operations statistics are reported continuously to the main computer and the screen displays the following:

- Number of workpieces produced in each cell.
- Number of workpieces on pallets in the central buffer and in each cell, and the extent to which they have been machined.
- Statistics over each type of stop.

To secure computer functions there are two computers working in the system, so that if one fails the other one will take over. This can be done only after manual updating of information, taking about ten minutes.

Advantages with FMS

Using computers to control and coordinate production in automated installations results in shorter lead-times and higher machine utilisation.

This means that the short retooling time of 20 minutes for a system like this can be produced in small batches and the production can be adapted to the needs of the assembly hall.

The following conclusions can be drawn for the FMS when compared with conventional solutions:

- Investments up to 20% higher.
- Tools costs from 10% to 20% lower.
- Personnel reduced by up to 50%.
- Floor space from +30% to -30%.
- Number of machines reduced by 50%.
- A cost reduction up to 25% per produced component.

Other improvements include:

- Higher quality.
- Fewer rejects.
- Reduced inspection costs.
- Less physically taxing work.
- Lower noise level.

The new labour implications of FMS

Through FMS the occupational structure will change. There is a labour-saving effect, workers are less tied to machines and are more independent of the production rhythm. It also means that the heaviest and dirtiest parts of the work disappear.

The people who will work in a new installation have to be chosen as early as possible so that they can take part in the designing of the working requirements and at the same time be given the necessary training. In this way machine operators are highly committed as they obtain the opportunity to take a greater responsibility for, among others, planning, programming, supervising of the machines and product quality. As a result, the skill and commitment of the operators increases and at the same time the company improves its competitivity and increases job security.

TECHNIQUES FOR AUTOMATIC IDENTIFICATION IN FMS INSTALLATIONS

J. Lindgren
Saab Automation AB, Sweden

In the future, when FMS installations are run with an unmanned third shift, the automatic identification of components, etc. will be one of the critical links in the chain. Different techniques and installation examples are presented in this paper. The techniques reviewed include barcode, microwave and machine vision.

Flexible manufacturing systems offer two major advantages:

- Short throughput times - Saab Automation has been involved with several installations which have cut throughput time from six weeks to several days.
- Profitable production in small series - Today's technology gives a tool for very rapid machine conversion, radically reducing the benefits of long-series production.

But FMS has been discussed mostly in terms of cutting operations. It has perhaps even greater applications on sheet-forming and cutting, welding and assembly.

In all of these, identification is a very important function. Automatic identification can be handled more effectively with FMS. Saab Automation has handled several FMS installations using different techniques for automatic identification. The different techniques are:

- Barcode.
- Microwave.
- Machine vision.

Barcode

An application example is at Saab Car Division. The company found a need to automate the production control system at its car body manufacturing plant. The previous manual reporting of work-in-progress, their location and status, was neither economical nor reliable. This condition was due to difficulties caused by a relatively free production flow in combination with several buffer stores in various areas of the plant.

The Saab EVS barcode reader was therefore chosen as the fundamental and most essential constituent when accomplishing the project. The system achieved the following results:

- Continuous reporting of plant and sub-divisions of production.
- High reliability and accessibility in reporting procedures and equipment, respectively.
- Late decision, regarding colour and car body type are allowed, which means greater flexibility and thus more pliability for variations in demand.
- Quick and continuous location of specific car bodies.

The production control system is built in modules. A method of construction which permits a simple and inexpensive expansion of the functions – to include among others:

- Supervision and control of the production flow in certain sub-sections and workstations.
- Supervision and control of components and subassembly provision for the correct car body, especially in the latter assembly stages.

Consequently, the system permits an expansion of the control tasks for further flexible production with continuous reliable output.

The production control system operates as follows. All car bodies are carried and transported through the plant on sledges with a fixed barcode plate attached to them. The codes are read in strategically placed reporting stations, each of which is equipped with a Saab barcode reading camera (Fig. 1).

The communication between these Saab cameras and a main computer is supervised and accommodated by two data-concentrators. Close to the passage of the first report station, i.e. just after loading the sledge with floor, roof and slides, the car body is assigned a specific code, which is used for location and defect diagnosis detection of the car body.

In the event of a communications breakdown with the main computer, up to a maximum of 3000 readings are stored in the data-concentrators, representing approximately ten hours of production. Additionally, each Saab barcode camera has a capacity for temporary storage of 50 readings, that is one hour of production.

Another example is the installation found at Luxor AB, a company manufacturing television sets, minicomputers and audio products. It has chosen Saab barcode reading cameras in its colour TV plant. There, the Saab EVS barcode reader forms the basis of a modern production control technology. The system contains five Saab EVS barcode readers together with an equal number of separate computer units and has

Fig. 1 A barcode reading system at Saab Car Division

various complementary subordinate equipment. The benefits of the system result in:

- Drastically reduced production costs.
- Generally improved product quality.
- Complete control of work-in-progress and thus shorter through-put time owing to a minimum demand of buffer stores.
- Increased production flexibility.
- Considerably improved ergonomical and contendable work situations.

Television sets are, because of their volume and physical build-up, well suited to line production. Luxor AB produces 1000 colour TV sets daily, divided into 30 different types. Consequently great demands are put upon the production control equipment, mainly because of the need for extensive control and adjustment procedures induced by this kind of free-flow production.

Microwave

In some installations it is necessary to have an identification media which needs to be changed during the manufacturing process. At the Saab Automation FMS installation at Seco Tools, the Philips PREMID system for identification and the Data Carrier system was used in the production cell for the tool holder. In this application the ID-plate is mounted underneath the pallet containing specific process information,

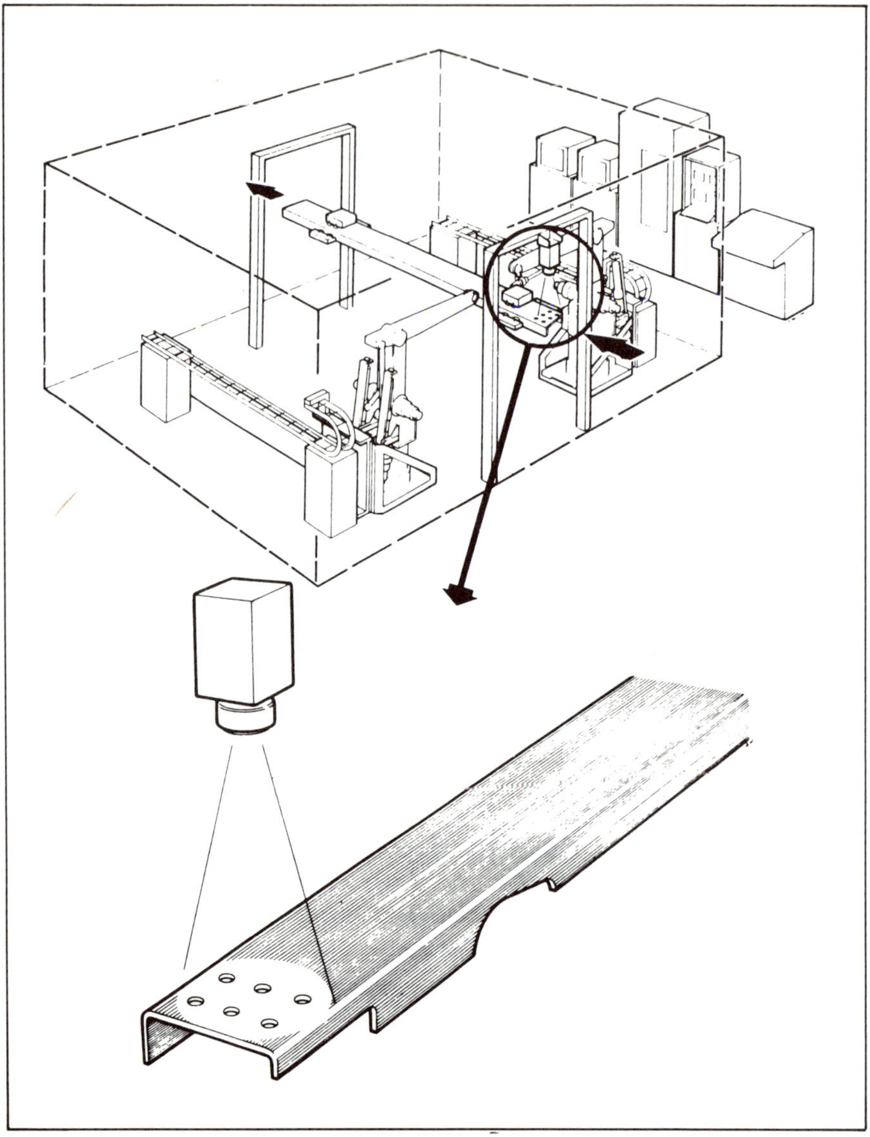

Fig. 2 A vision system at Volvo identifies chassis types for cutting

such as the number of pieces on the pallet and the program number to be used for different manufacturing operations by the robot and machines. When all pieces on the pallet have been processed an 'OK' message is programmed into the ID-plate.

Machine vision

In many cases it may be difficult to apply a barcode reader or a microwave system. Then it is necessary to have an image handling system in order to see other parameters and thus automatically identify various types of product. Typical examples of installations, in which artificial vision has allowed highly rational and flexible flows of materials and processes, are found at:

- Norsk Wallboard - an EVS 300 system is used to identify different boards.
- Electrolux (Mariestad) - six systems perform various identification tasks, including 'seeing' the pattern of holes on the back of a refrigerator.
- Volvo Trucks - an EVS 300 system is used to identify chassis sidemembers in the plasma cutting cell.

The operation requires the stamping of a 'hole-picture' in each sidemember. This hole-picture is the identity of the sidemember. The sidemember, or rather two, three or four sidemembers at a time, is transported to the robot cell. Before it enters the actual robot cell, an operator will enter a code into the computer for that specific package. The code decides how the robot should cut.

The sidemember enters the robot cell in order to be cut at the rear end. After that it moves slowly forward until the vision system has identified the hole-picture. The camera gives the information to the main computer, which selects the right cutting program for the robots. After cutting, the sidemembers automatically move out from the cell. Once the operator has entered the code, the process is fully automatic (Fig. 2).

AN FMS FOR MECHANICAL COMPONENTS

P. Vicentini and A. Gambini
Alfa Romeo Auto SpA, Italy

A feasibility study and project were carried out to verify technical and economical conditions under which it is possible to develop an FMS to manufacture prototypes and special short-run engines and mechanical components. The system has a modular configuration and is compatible with further development. The FMS handles prismatic and rotational components. It includes machining centres, a lathe, a grinding machine, washing machines, a measurement unit, and is controlled by a hierarchical computer system. The feasibility study was developed using simulation techniques for performance evaluation and optimisation. In this FMS it is possible to integrate new technology and to have better production autonomy.

Tough competition in the car industry is caused not only by a production capacity that exceeds market demand, but by the fact that, at least from the consumer's point of view, the car is in many ways a mature product.

In order to survive, car manufacturers must identify key market demands. They will not be created by the consumer or by economic factors.

It is therefore the designer who must use skill and imagination to identify the design and performance of the car which will be most successful in the eyes of the consumer.

However, the speed with which an idea can become a prototype, then a mass-produced prototype and finally an implemented production line, is more and more important in helping to tighten the interactive ring between the proposed car and the consumer's acceptance, in order to guarantee the manufacturer's survival.

Besides demanding considerable time and human resources, the complicated operations forming the main phase of a new car programme also require increased investments and therefore are expensive. It is essential that each car manufacturer analyses all the opportunities that new technologies have to offer to make this phase of the industrial process more effective and efficient.

From this point of view, the 'car' product is not mature, but one that is rapidly developing under the influence of modern technologies which will change not only the product itself, but, in particular, its creation, design and construction.

Prototypes in the car industry

The manufacturing processes involved in the construction of a new model are extremely complicated. Not surprisingly, the complexity of this phase is closely linked with the type of car to be built and the future production capacity foreseen.

There are several different types of prototypes including:

- Competition car prototype.
- Research car prototype.
- Medium/mass-produced car prototype.
- Prototype of car produced in limited and special series.

Each prototype is associated with build construction phases, costs, times, and different degrees of risk. Leaving competition prototypes aside, as the manufacturing process factor is seldom not involved, at least directly, this leaves the three remaining prototypes.

One or two prototypes, also known as 'concept cars' are built over a period of three to four years. The main target is to check the feasibility of certain new technical alternative solutions, particularly as regards car performances. The research prototype generally is not subsequently built in production in its creative state, but its technological solutions have 'spin-off' potential for production vehicles.

The prototypes for medium/mass-produced cars are divided into two phases. The function prototypes are used to check performances; while the industrial prototypes are used to check industrial feasibility and recheck performances.

The first are built with basic equipment, generally in a few models to be used for different performance tests. The second are constructed partly with production equipment and partly with equipment to simulate the production equipment. These vary according to the model and degree of innovation to be used for the production process and the performance tests.

The final case is the 'special' prototype for short runs. These prototypes range between functional and industrial versions. The costs involved make it difficult to balance performance and industrial requirements with the number of cars to be produced in order that the overall operation produces a profit.

This is a difficult problem, but modern technologies can lead to considerable benefits, particularly in the case of prototypes for short run 'specials'. In particular, flexible manufacturing systems (FMS) permit high performance manufacture at fairly low costs and at reduced times.

An FMS for prototype engines and transmissions

The FMS studied by Alfa Romeo has a number of different objectives, some of which affect production, whereas others are directed to research. There are several different targets within the objectives which have to be met:

- To develop prototype engines that assist in the design of new products.
- To develop various prototype engines to carry out a large number of trials such as durability, reliability and fuel economy.
- To develop a series of special engines to be produced in small numbers for sports cars requiring homologation.

Amongst the objectives with a research character are:

- The use of the FMS as a pilot plant to study new techniques for production; to verify and value the fields of applications, the economic advantages, and the ability to integrate with other traditional technologies; and to study the possibility of introducing innovative techniques in the production of traditional high-volume engines and associated components.
- The use of the FMS as a field of study to provide new professional information.

The feasibility study was carried out within a research contract between Alfa Romeo and CNR (National Research Council) for the 'finalised project of mechanical technologies'.

The product

The plant was studied to use two different families of components: components of prismatic shape and components of rotational shape.

In the prismatic family, components include all mechanical units with dimensions falling within an 800mm cube of made in cast iron or aluminium and requiring cutting, milling, boring and thread-cutting along axes parallel to the axes of the machine.

Among rotational components to be machined in the FMS are all those mechanical components in steel falling within a cylinder with a length of 1400mm and a diameter of 300mm requiring turning or grinding along the parallel axes, as well as boring and thread-cutting along any axis. In addition to these operations it is planned to study the possibility of inserting heat treatment using impulsed radio frequency.

The items intended as families for the FMS include engine components and the transmission. In particular the following components have been chosen:

Engine (prismatic)	Gearbox (prismatic)	Miscellaneous (rotational)
Engine cylinder head	Gearbox	Crankshaft
Engine cylinder block	Gearbox cover	Camshaft
Cylinder head cover		Gearbox shaft
Oil sump		
Cover		
Inlet manifold		
Drive shaft support		
Distributor shaft support		

The proposed engines are of three types: V6; four-cylinder in-line; and opposed cylinder (boxer). The development and design of the FMS has been subdivided into various subsystems, each devoted to a specific function. The subsystems which constitute the FMS are:

- The subsystem of the production process.
- The subsystem of the mechanical work.
- The subsystem of the assembly devices.
- The subsystem of the handling devices.
- The subsystem of the tools.
- The subsystem of the pallets.
- The subsystem of the control computers of the FMS.
- The original subsystem of the auxiliary services.

Functional architecture

The control system has a functional architecture based on three hierarchical levels of control, each of which has well defined aims, functions and frequency operations.

The system has a direct action or 'control' which works a series of actuators, measurement or monitoring, and sensors. These provide a series of information from the physical plant which is used by various levels of control to create a reprogram.

The control system has an external input constituted by the nominal productive requests which define what to produce, in what quantity, within what date, and by an external output constituted of reports, statistics, diagrams, and so on, to various levels in detail.

The physical Flexible Manufacturing Systems will have the raw material as input and the finished product as output.

Its logical function structure is illustrated in Fig. 1. They distinguish two lines, typical of every automatic feed-back system:

- A control line or 'direct-chain'.
- A monitoring line or 'chain of measurement'.

The line of control has been divided into three hierarchical levels on the basis of frequency within which these operate.

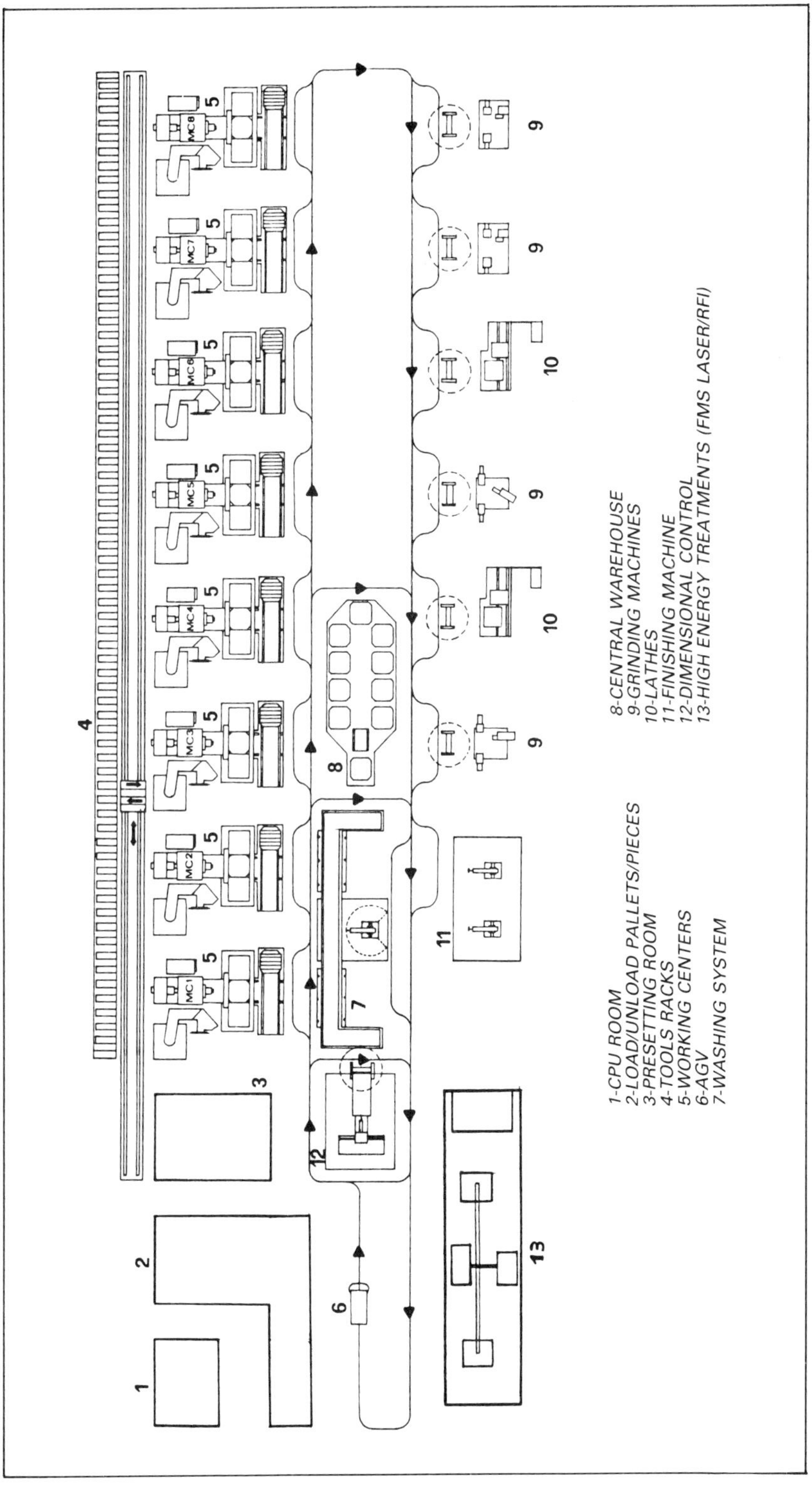

Fig.1 Layout of FMS

- Strategic level - frequencies of monthly operation (but it can be varied by the operator).
- Tactical level - frequency of daily operation.
- Local level - frequency of operation given by the dynamics of the events in the FMS.

The line of monitoring is the 'chain of measurement' of the FMS physical system; from this it detects by means of 'sensors' the performances.

It is divided into two levels:

- Monitoring of processing - Which collects all the informations from the system and within a short time gives the information to local level to correct the actions.
- Monitoring of system - Which filters the information from the monitoring process and associates them in a way to supply a whole vision from the work done totally supplied by the system.

Simulation

During the development of the project, especially for the parts concerning optimisation of the layout, simulation programs of the analytical type were used.

The programs used are: SELECT; GLOBAL; BATCH; and QUICK.

The first three carry out their functions during the period of the design of the FMS, while the last is the proper simulation program.

- SELECT - This outlines the minimum investment of the operating units to achieve the required production. The algorithms on which such programs were based are integer linear programming and 'branch and bound'.

- GLOBAL - This determines ahead of a specified plan of production the minimum time required by the plant, previously defined, to complete the production; in any case there are no constraints of any kind to the simultaneous machining of points in the plant.

- BATCH - This has the same function as GLOBAL but operates in case of constraints (that are to be specified) to the simultaneous presence of the workpieces. Typical constraints include those imposed by the limited capacity of the tool storage of the machine. The program schedules the plan of production into different sub-periods including just those workable pieces handled simultaneously in accordance with the imposed constraints. The total completion time of the plan of production estimated, has the limitations worked out and it will be defined by GLOBAL which represents the absolute minimum.

- QUICK - This is the program that simulates the steady-state running of the system of the plan. It supplies various information that, opportunely interpreted, allows introduction of layout modifications to optimise the productive rate of the system.

Other aspects of the simulation program include:

- BEST-LOADING - Bearing in mind different technological alternatives and routes available FMS determines what may be the best distribution of the pieces among the machines.
- BEST WIP - This gives the number of pallets that must be constantly in the system (work-in-progress) to assure maximum production rate. The WIP is shared for pallet type.
- UTILISATION - For every operation unit this gives the number of pieces waiting in the FMS queue, the average waiting time and the flow of pieces.
- PRODUCTION - The production obtainable is function of number of pallets placed in the system. The production is supplied under the form of a curve of production/pallet. This moves asymptotically to the maximum value of production obtainable. The number of pallets corresponding to maximum (asymptote) is the best WIP as defined previously.

Concluding remarks

The FMS as previously defined may produce theoretically about 5000 kits of engines and mechanical groups on the basis of two shifts a day.

As many new problems are involved, prudence suggests a step-by-step approach, not only because of gaining more know-how but because the financial aspects are considerable.

So we are carrying out, therefore, a realisation of a first, reduced part of the plant, that has functionally all the characteristics of an FMS research plant.

As part of another contract with CNR, Alfa Romeo is studying the possibility of introducing on-line FMS-Laser station for heat-treatment, continuous welding of mechanical parts or a pulsed radio frequency station for heat-treatment of crankshafts. (An FMS-Laser system was patented by Alfa Romeo and CISE, an Italian Research Institute for Development of Laser Systems.)

References

[1] Bassi, A. 1984. From the first line to the prototype. In, *ISATA '84 Conf. Proc.*

[2] Franchetti, I. and Grisoni, L. 1984. A flexible laser system provided with a laser beam composition unit for industrial application. In, *ISATA '84 Conf. Proc.*

[3] Bott, H. Etwicklungsziele bei Rennwagen, *Sportwagen und Limousinen.*

[4] Appel, H. and Hilber, K.P. Problematik der Abschatzung von Entwicklungslinien fur die product und produktionstechnolgien Zukunftiger Personenkraftwagen.

[5] Hartley, J. 1984. *FMS at Work*. IFS (Publications) Ltd, Bedford, UK.
[6] Hayes-Roth, F., Waterman, D.A. and Lenat, D.B. 1983. *Building Expert Systems*. Addison Wesley.
[7] Snowdon, J.M. 1984. Planning vehicle design. In, *Proc. Int. Cong. of Vehicle Design*, Vol. 5, N.1/2.
[8] De Luca, A. 1983. Computer aided techniques for FMS design and management, a case study. In, *Proc. 2nd Int. Conf. on Flexible Manufacturing Systems*, pp. 605-614. IFS (Publications) Ltd, Bedford, UK.
[9] De Luca, A. and Seco Giardo, G.M. 1982. Progretto assi emiro da calcolatore di sistmi versatili di lavorazione. *PIXEL*, 4.

6

WAREHOUSING, AGV SYSTEMS AND SIMULATION

The concept of just-in-time manufacture might infer that, at a stroke, warehousing and advanced material handling techniques will diminish. On the contrary, even if the end-producer can manage to reduce inventories significantly, the component supplier has its own problems of developing technologies in the area of inventory control and material handling. Warehouses may become smaller but they will be used more intensively and they will have to handle a greater variety of components. Meanwhile, the inter-relation of the production line, the warehouse and the assembly line will continue to raise problems, the solution of some of which can be achieved only through effective and intelligent use of simulation techniques.

AUTOMATED STORAGE AND TRANSPORT OF STAMPINGS

L. Dittrich
Schindler-Digitron, Switzerland

In May 1983 a high-bay warehouse (HBW) for pressed automobile body parts went into operation at the Sindelfingen plant of Daimler-Benz. This HBW system is part of an extensive transport installation. It serves as a buffer store between the press shop operating in batch production and the continuously operating assembly lines. This paper describes the two separate implementation phases.

Detailed planning and realisation of the high-bay warehouse (HBW) system was carried out by Schindler-Digitron. As basic planning data Schindler-Digitron was provided with a study on the feasibility and optimum location of the HBW within the Daimler-Benz plant area (carried out by the Fraunhofer Institute (IPA) of Stuttgart) together with Schindler-Digitron's own study on an automated pressed component transport system (PCTS). Due to the complexity of the system it was divided into two separate phases for implementation.

Phase 1 – Storage system (HBW)

Scope

The order for the first phase comprised a high-bay warehouse - a self-supporting steel-framed structure; automatic stacker cranes; transport to the identification points and in the HBW; and warehouse administration and control computers for the entire system.

The client undertook responsibility for the floor surfacing, roof and wall linings, and the internal supplies and services.

Objectives

The new HBW (Fig. 1) had to fulfil the following needs: replace the external stores which were located up to 50km away from the plant; reduce material handling outside the plant; optimise location of the

Fig. 1 Front view of high-bay warehouse (HBW)

HBW for material flow within the Daimler-Benz works area; quality-conform transport and storage of pressed components; reduced intermediate storage within the plant; achieve a high degree of transport automation within the HBW allowing connection to the PCTS at a later date; and offer computerised stock-keeping.

Requirements

Every day (14.66 hours in two shifts) 2000 steel pallets with stamped or pressed components have to be stored in the HBW, and the same number retrieved. The HBW is supplied from three independent press shops, a stamping shop and a plant for pressed/welded components.

The components are delivered to the warehouse in 35 different pallets, divided into pallet types 1 and 2. The maximum weight is 1500kg for pallet type 1 and 2000kg for pallet type 2. Pallet storage within the HBW is also separated into pallet types 1 and 2. In the vicinity of the HBW weight and data records are made for each pallet, and after identification pallets are followed automatically by the data retrieval system until dispatch to the body shop.

The HBW supplies the three assembly halls of the body shop with pressed components, involving various destinations within each hall. The transport system in the HBW approach zone had to be designed to conform with the predetermined construction limits.

Additional transport aids such as auxiliary pallets and so on, had to be avoided, especially bearing in mind the PCTS . An important requirement was the possibility of loading and unloading the transport system with fork-lift trucks. Structural alterations to the pallets such as welding on transport skids were undesirable, because although this facilitates pallet movement within the transport system it hinders the collection and delivery of pallets by fork-lift trucks in the production zone.

Due to the high rate of throughput the transport system had to have adequate buffer facilities, especially in the type 1 pallet zone. Storage and retrieval of the pallets takes place flush with the floor at elevation ± 0 (ground floor). At a later date connection will be made between the HBW and the PCTS at basement level (elevation -5.5m), and the necessary structural measures had to be taken into account during the construction of the HBW.

System overview

Detailed planning of the HBW had to take account of the conveyors and transport systems at basement level as well as transport installation on the ground floor. For determination of floor area requirements in the basement it was assumed that an automated guided vehicle (AGV) transport system would be employed.

The new HBW requires about 9000m^2 of storage area including stacker crane space requirements, and about 1275m^2 surface area for the transport system in the warehouse approach zone. In comparison with this, about 35,000m^2 was required for block storage in the previous external warehouses.

The HBW comprises the following sections:

- A steel-framed HBW building ($L \times B \times H = 106 \times 85 \times 85$m) with a total of 12 aisles and 24,600 storage compartments, divided into three aisles with 9600 compartments for pallet type 1 and 9 aisles with 15,000 compartments for pallet type 2.
- Three automatic stacker cranes with two independent telescopic forks for pallet type 1, maximum load 3000kg.
- Nine automatic stacker cranes with independent telescopic forks for pallet type 2, maximum load 2000kg.
- Three transport routes each for storage and retrieval with transfer

cars equipped with lifting table and four buffer places for pallet type 1.

- Nine transport routes each for storage and retrieval with transfer cars equipped with lifting table and three buffer zones with pallet type 2.
- Two identification points with built-in weighing scales, profile check and transport system with one transfer car equipped with lifting table for pallet type 1.
- Two identification points with built-in weighing scales, profile check and transport system with two transfer cars equipped with lifting table, for pallet type 2.

System operation

All pressed component pallets are collected from the press shop either by fork-lift truck, or on trailer trains under the roofed-in approach to the HBW, and then placed in intermediate storage. In the roofed-in approach zone the pallets are sorted by fork-lift trucks (which operate only in this zone) into types 1 and 2 and placed in readiness on the identification point transport routes (Fig. 2). Light signals inform the

Fig. 2 Set-up place identification point

fork-lift truck drivers whether a pallet is load-conform or not. After the signal 'load-conform' has been received the fork-lift truck driver initiates transport of the pallet to the identification point by operating a manual switch. Transfer cars are used for transporting the pallets to the identification point. At the identification point the weight of the pallet is determined on scales and reported on-line to the warehouse computer. At the same time the exterior dimensions of the pallet and the clearance for the telescopic forks of the stacker crane are checked.

The operator at the identification point must read in the barcode data on the card accompanying the material, and, if necessary input further data as the pallet number through a terminal.

After all checks and data inputs have been completed the pallet is transported by transfer car to the identification delivery point which is situated in the HBW approach zone. Light signals inform the fork-lift truck driver in the HBW approach zone if a pallet is ready for collection. In this zone the fork-lift trucks are equipped with an automatically readable identification system, and as soon as the pallet is collected from the delivery point the fork-lift truck number is automatically read (Fig. 3). The driver is informed of the pallet

Fig. 3 Fork-lift truck collection point (pallet is ready for collection)

destination address in the form of the stacker crane aisle number by a large digital display (Fig. 4).

On arrival at the stacker crane aisle, the fork-lift truck deposits the pallet at the deposit station and the driver is informed by light signals whether the pallet has been correctly deposited (pallet 'conform'). If the pallet has been deposited at the wrong aisle a large digital display is given of the correct address.

A transfer car then transfers the pallet from the fork-lift truck deposit station to the stacker crane collection point. During transport the pallet passes again through a profile check in order to establish any displacement of the load during transport by the fork-lift truck. On arrival at the collection point, the pallet is picked up by the telescopic forks of the stacker crane and transported to the compartment address predetermined by the computer. The type 1 stacker crane normally takes two pallets. Storage in the HBW is either in two directly neighbouring compartments or at two different addresses. Type 1 pallets are stored

Fig. 4 The fork-lift truck has collected the pallet (the pallet destination address is shown by digital display)

supported on their feet, whereas type 2 pallets are stored on special profiles built into the compartment because of their varying floor height.

Pallets are retrieved in the opposite sequence to storage. The stacker crane places the pallet on the crane delivery point, and from here it is transported by transfer car to the fork-lift truck collection point (Fig. 5). Light signals indicate that a pallet is ready for collection. As soon as a fork-lift truck has collected the pallet its destination address is indicated by a large display. The fork-lift either transports the pallet directly to the washing plant or to one of the three trailer trains going to the body shop, situated under the roof of the new HBW approach zone.

The entire system is administered and controlled by two PDP-11/44 computers with a third computer as standby in case one of the active computers breaks down. The administration computer is responsible for material scheduling tracking and stock-keeping. It receives information about identification points and processes on-line orders for pallets to be retrieved and sent to the production areas.

The control computer controls and optimises the physical movements of pallets by transfer cars, stacker cranes and fork-lift truck in response to commands from the administration computer. It also keeps a record of free storage locations and assigns locations to pallets entering storage.

Phase 2 – AGV transport system (PCTS)

During construction of the HBW, planning of the PCTS was carried out. In the area of the HBW delivery point structural measures in the

Fig. 5 Fork-lift truck pick-up/deposit stations in front of HBW

basement were completed on the basis of the feasibility study. The remaining transport routes were not yet defined. The transport means proposed were an automated guided vehicle (AGV) transport system or an overhead monorail (OM) system.

Requirements

Based on the varying production requirements, the flow of material can be divided into two circuits, one for empty pallets and the other for full pallets. The new PCTS (fig. 6) has to fulfil the following transport needs.

- Full pallets with pressed components from the press shop via the HBW to the body shops.
- Empty pallets from the body shops via the empty pallet transfer or reloading station to the press shop.
- Other components from the delivery point to the receivers, and the empty pallets back to the delivery point.

As far as possible all transport between the press shop and the body shops should be carried out directly (only bypassed to the warehouse or transfer station). Transport within the body shops in connection with workshops external to Daimler-Benz should also be carried out by the PCTS. The PCTS must be able to take over pallets in the basement which have been delivered on the ground floor, and transport them to their destinations.

With the exception of the AGV system, all parts of the transport system (elevators, storage/retrieval conveyors, and so on) must have reserve capacity of 20% above the material flow. The controls of the automated transport system must be integrated into the production sequence control of the pressing works - pressed component warehouse - body shops.

In particular the following requirements must be fulfilled:

- Consumer supplies must be assured at all times.
- Transport times (operating times) must be kept to an economical minimum.
- Early registration of all pallets must be ensured by 'advanced' identification points.
- The controls of the automated transport system must be integrated into the pressed component control and administration system.
- Operation must be possible in the bypass mode, taking account of the state of alteration of components and other influences.

The automatic transport system must be flexible regarding material flow, number and position of transfer stations. In particular it must be possible to relocate production sectors into other parts of the assembly halls.

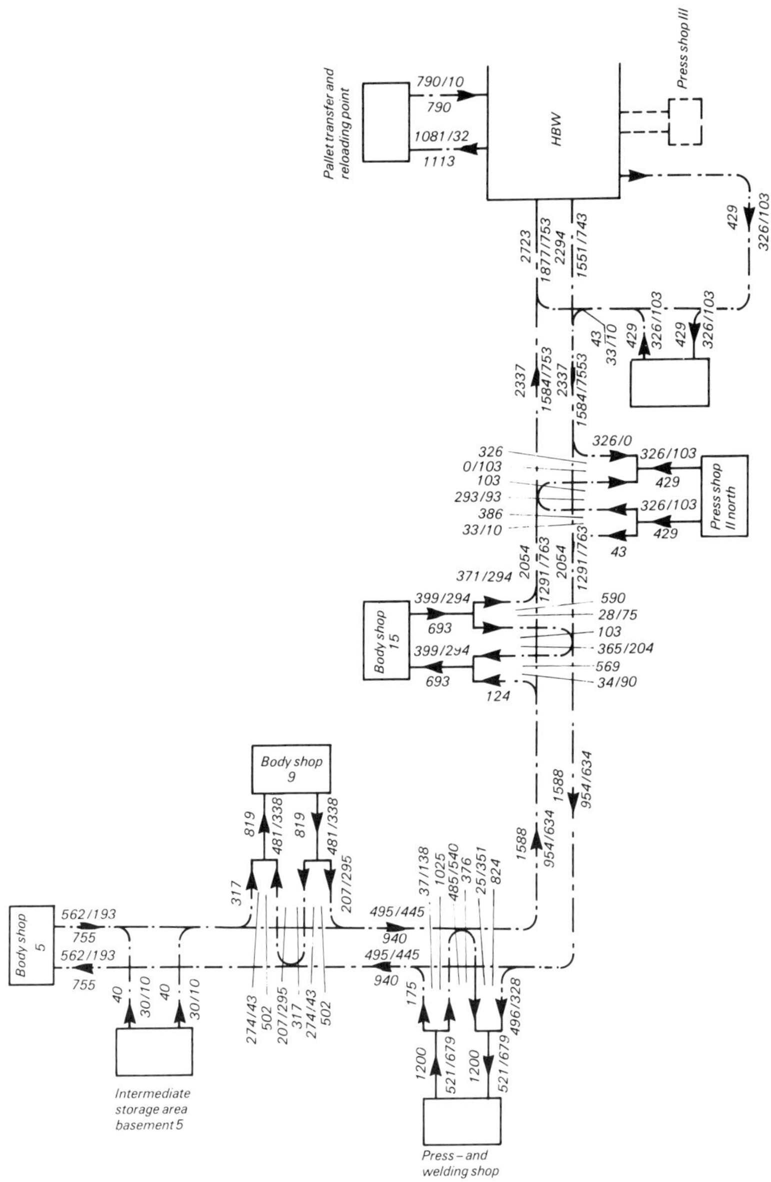

Fig. 6 Material flow AGV transport system (PCTS)

Planning results

Before determining the transport vehicle routes within the production area, it was necessary to select the vehicles themselves. The two systems, AGV and OM, were compared based on the client's evaluation criteria. It was decided finally to install an AGV system with inductively guided robovehicles.

Based on the pallet dimensions (pallet type 1 = 1000 × 1200mm, pallet type 2 = 1700 × 2400mm) an investigation was first carried out to determine whether one robovehicle could transport two type 1 pallets or one type 2 pallet. However material flow requirements allowed only about 50% of the robovehicles to be simultaneously loaded with two type 1 pallets. Furthermore it was not sure if these pallets would be transported to the same destinations, and subsequent planning was therefore based on the use of two different types of robovehicles. After the transport system had been selected, the transport routes for the AGV were examined.

In the HBW approach zone structural work had already been completed. HBW storage and retrieval takes place on the ground floor at elevation–5.5 m. In the production area an investigation was carried out to determine whether the existing ground floor transport routes at elevation 0.0m could be integrated into the system.

Transport density computations showed that at peak periods vehicles would pass certain crossings at intervals of 20s in both directions. This would have precluded the possibility of using the transport routes within the production area running transverse to the direction of travel of the AGV.

One of the client's requirements however was that transport within the production area should not be affected. For this reason separate transport routes were arranged also for the AGV basement level, elevation -5.5m. At central points within the production area the AGV is transferred by elevator to the ground floor, elevation 0.0m, to cocllect or deliver pallets.

Another critical point is the robovehicle power supply. The client rejected the idea of changing batteries or a special charging zone in the basement beside the AGV transport routes. Battery changing would have required an extensive battery charging room and additional personnel at change of shift. Recharging batteries on the AGV within a special charging zone would have required additional vehicles.

In order to fulfil the client's requirements a new power supply concept had to be worked out, as a result of which the AGVs have been equipped with nickel-cadmium (Ni-Cd) batteries. Recharging current is taken from power supply rails during transport or at a special disposition stops in the station zones. The power supply rails are arranged inside the tunnels connecting each production hall. The Ni-Cd batteries can be recharged at very high current, so that the energy consumed during each journey is automatically replenished. A special disposition program allocates the robovehicles to route sectors with power supply rails after the late shift, for recharging during the night.

The conveyor system in the basement is identical to that on the ground floor. The conveyor systems in the basement are loaded by telescopic fork as a connecting link between the conveyor system and robovehicles.

System overview

The PCTS is made up of the following system components:

- Three stations with dual elevators and integrated identification points for pallet types 1 and 2.
- Four stations with dual elevators in the body shop zones.
- One special station with dual elevator for transferring empty pallets.
- Three conveyor systems for storage/retrieval of type 1 pallets and nine for type 2 pallets in the basement of the HBW approach zone.
- Approximately 4000m of active transport route for robovehicles.
- 25 robovehicles for transporting type 1 pallets.
- 60 robovehicles for transporting type 2 pallets.

Operational sequence

The pressings are collected by fork-lift truck from the pressing line and brought to the delivery point in the station zone (Fig. 7), separated into type 1 and type 2 pallets. The corresponding AGV collects the pallet by travelling underneath it and raising its platform, and transports it to the identification point (Figs. 8 and 9).

The following functions are carried out at the identification point: weighing of pallets; determination of profile; identification of the contents; and a check or determination of the number of components per pallet.

Before identification, each pallet is weighed and its profile is determined. Both the weight and the profile data are automatically registered and processed by the control system. The article number (for identification of the pressed components) is read in barcode from the material identification card by the identification point operator (Fig. 10). The quantity of large components is checked visually by the operator, and for bulk goods the computer determines the quantity based on the weight.

After identification has been completed, the AGV system receives a transport order in accordance with one of the following three categories:

Storage in the HBW. Identification has been successfully completed. The pallet has to be stored in the HBW. The transport order included the destination HBW and the pallet identification. The AGV transports the pallet to the warehouse delivery point, and it is halted at the transfer station to the storage conveyor of the preselected warehouse aisle. Here the pallet is unloaded by an automated telescopic fork device. The corresponding storage bay data is transmitted to the warehouse system,

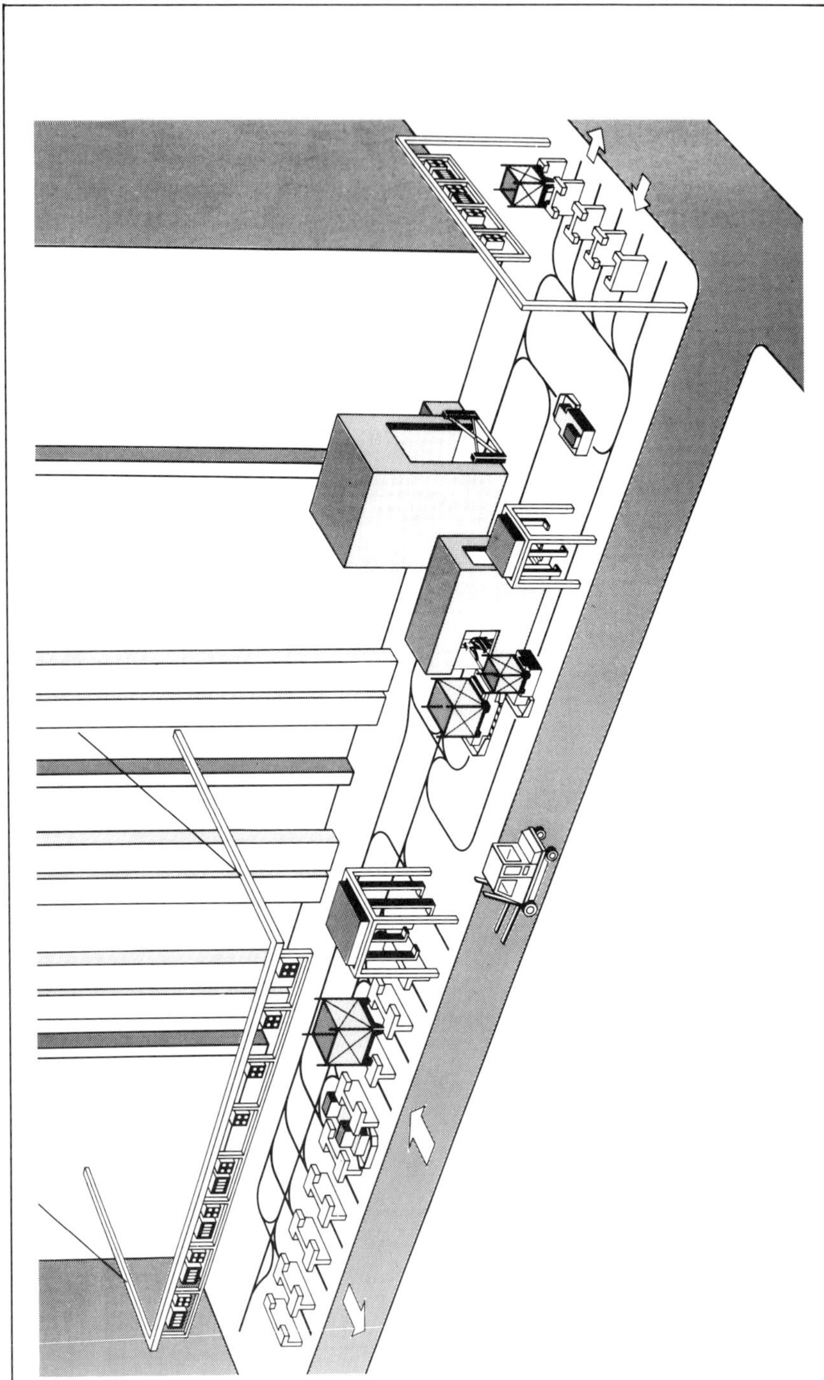

Fig. 7 Station at press shop area

Fig. 8 AGV for transporting type 2 pallets

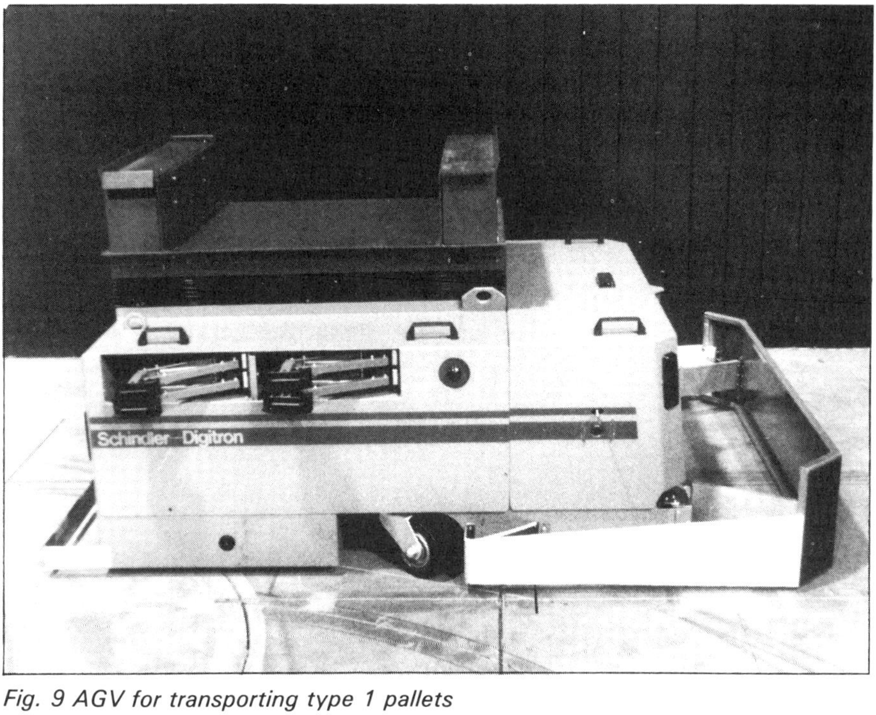

Fig. 9 AGV for transporting type 1 pallets

Fig. 10 Operator room showing schematic display of the HBW

which subsequently completes the storage procedures with the storage conveyor and stacker crane.

Bypass. Here too identification has been successfully completed. However an order has just been received for a pallet with this particular pressed component. For this reason the pallet is not stored but transported directly to the 'consumer' destination where it is required.

The destination 'consumer' describes a definite station to the supervisory system, which then guides the AGV with its pallet to the station and searches there for a free delivery point. The pallet is transported from this delivery point to the consumer by fork-lift truck in a later phase via the fine-distribution system.

Rejection. Pallets can be rejected at the identification point, for reasons such as negative profile check results, excess weight, missing material identification card, and so on.

The identification point operator notes the reason for rejection on a card which he/she attaches to the pallet. The pallet is then transported to a predetermined delivery point in the station area.

Retreival from the HBW

The number of pressings required for the assembly lines is determined by dispatchers and ordered through the central order registration system. In the final stage the pallets are retrieved automatically.

The warehouse system determines the pallet to be retrieved and transports it by stacker crane and retrieval conveyor to the transfer station for the robovehicle transport system (Fig. 11).

The telescopic fork device collects the pallet from the transfer station, while the control system allocates an empty AGV to the corresponding aisle. The telescopic forks place the pallet on the AGV, which remains inactive during the transfer procedure. The AGV subsequently travels to the destination station determined for the transport order. At this

Fig. 11 Input/output transfer cars to HBW

station the control system selects a free transfer point and guides the robovehicle to it. By lowering its platform the robovehicle places the pallet on the transfer point.

Subsequent transport to the consumer is carried out by fork-lift trucks.

Transporting empty pallets

Empty pallets occur on the assembly lines, and are required in the press shop.

Type 1 pallets cannot be altered, but type 2 pallets can be fitted with various mounting devices according to the pressed component.

Empty pallet requirements. For pressings, the empty pallet requirements per component are determined by the stamping works computer over the entire time period in which the component is to be pressed, and transmitted to the PCTS. While the presses are being set up for a particular component, a definite number of empty pallets must be placed in readiness. Afterwards further pallets are supplied at intervals. Each time a pallet with this particular component is recognised at the identification point another empty pallet must be ordered.

Transport of empty pallets. The AGV generally transports empty type 2 pallets delivered at a body shop station to the empty pallet transfer station. If however the AGV collects a type 1 pallet from a body shop station, the control system must check whether there is any requirement from a press shop for this type of pallet. If this is the case the AGV transports the pallet directly to the press shop. Otherwise empty type 1 pallets are also transported to the empty pallet transfer station. Each delivery point in the press shop stations is equipped with a large 4-digit display. When a pallet is delivered the consumer identification (pressing line, press number) is displayed.

The PCTS is controlled by two PDP-11/44 computers. Since the same type of computer is used as in the HBW, there is no need for additional back-up computers.

AUTOMATED WAREHOUSE FOR CAR STORAGE

Y. Matsuyama
Ishikawajima-Harima Heavy Industries Co. Ltd (IHI), Japan

An automated warehouse for car storage was developed in 1977, and now six plants have been constructed in Japan. From the stand-points of storage volume and throughput, system layouts can be classified in three patterns. In 1984 the newest one was handed over to Nissan Motor Co. Ltd to be used exclusively for temporary storage and marshalling at an export wharf. In this project, various advanced techniques were applied to realise the required upgraded capabilities, such as system simulation, seismic analysis, enhanced control functions, and so on. This paper helps to share with other prospective users IHI's experiences about technology and usage.

In line with the rapid growth of the economy, the Japanese automobile industry now ranks in second position for the number of cars assembled in the world. Many efforts have been made in the rationalisation of the distribution system to cope with such a rapid growth in demand, in addition to the expansion of production facilities.

Through the cooperation of Nissan Motor Co. Ltd, one of the largest automobile manufacturers in the world, IHI has developed a high-rise automated warehouse system for passenger cars, replacing the conventional outdoor parking lot or multi-storey garage, and completed the first plant capable of storing 5000 units in 1977. Since then six plants have been constructed, and thus IHI has attacked the problem of rationalising the distribution system from the aspects of improving the storage efficiency and systemisation of vehicle handling operations.

The distribution of cars can be divided roughly into two areas: domestic sales and exports, and some kind of storage facility is always installed at each node of distribution (Fig. 1). Because of the high storage efficiency owing to its scale, automated car storage systems have been introduced mainly for the parking lots in assembly plants. As its effectiveness is proven and technological innovations have been made, automated systems have been adopted gradually by dealers in local cities with whom such a system had an obstacle in terms of economical

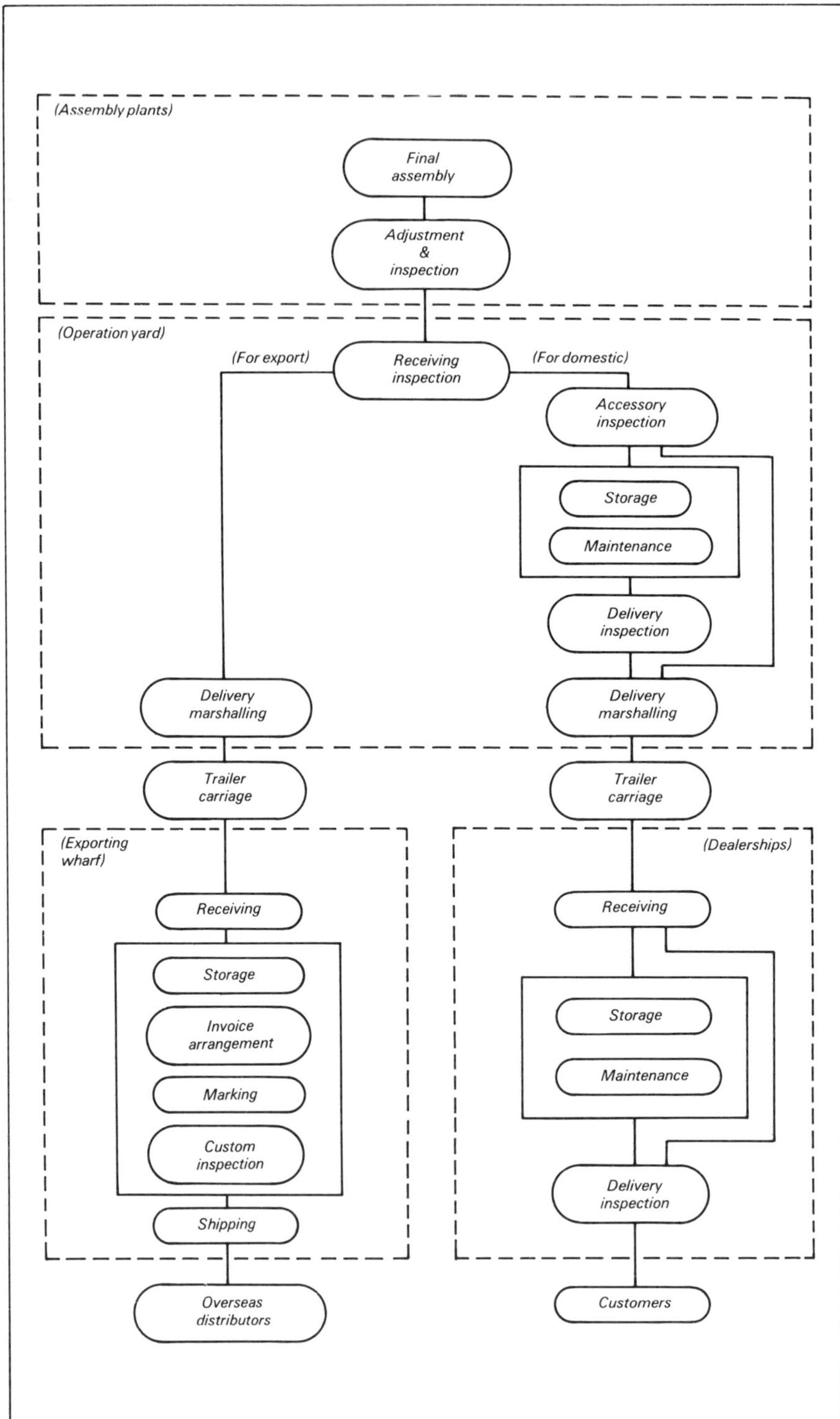

Fig. 1 Typical distribution flow diagram

System classification	Typical layout	Features	Proper capability	
			Storage volume	Throughput
Pattern I Compact type	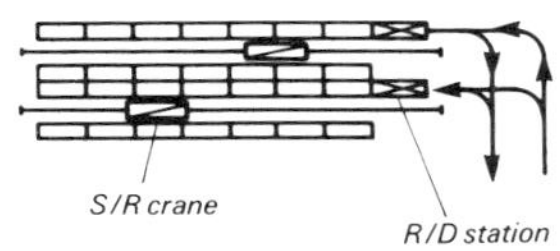	○ Simple & compact ○ Economical ○ Proper to small scale use as dealers or service centres	500 units ∫ 3,000	50-100 cars/hr (10 cars/crane/hr)
Pattern II General type	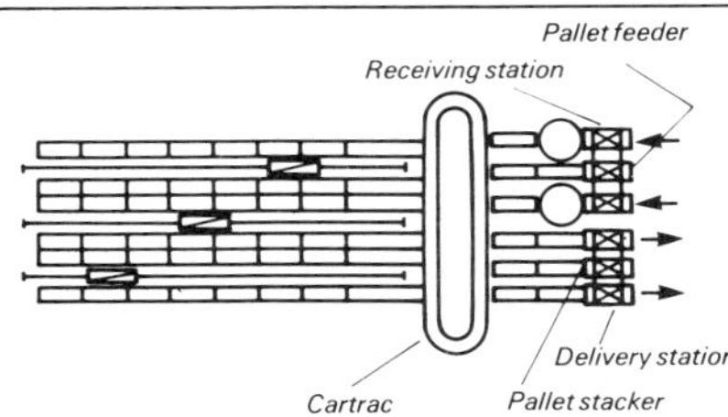	○ Effective for intensive operation ○ High throughput ○ Proper to large scale use as operation yards or exporting wharfs	5,000 units ∫ 20,000	○ Dual handling 200 + 200 cars/hr ○ Receiving or delivery only Max 450 cars/hr
Pattern III Extended type	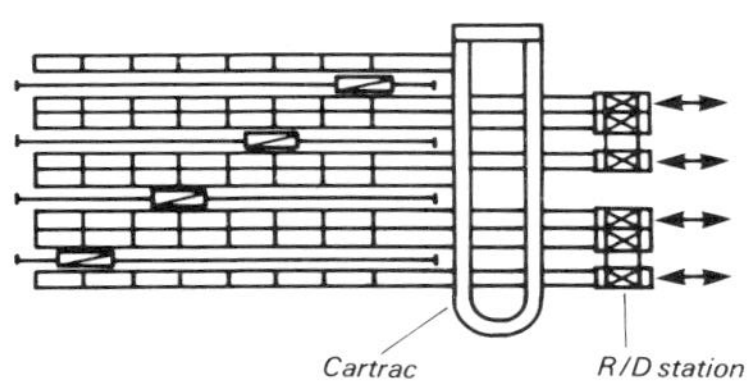	○ Cope with most high throughput ○ Proper to large scale and fluctuated operation, exclusively for exporting wharfs	10,000 units ∫ 30,000	○ Dual handling 250 + 250 cars/hr ○ Receiving or delivery only max 450 cars/hr

Fig. 2 System classification and application

evaluation. At the same time, they have been used for the yards of export wharves where the needs of large storage volume and massive throughput could not be satisfied concurrently.

Since the handling unit of the automated warehouses for cars is a mono-load, it does not require complicated processes like picking and inventory control of separated pallets. So, the main factors determining layout are storage volume and throughput capability. Accordingly, the most important points in the design are simplification and increasing overall performance. Automated car storage systems can be classified into the following three patterns based on past records and engineering data (Fig. 2):

- Pattern I - An independent receiving/delivery station operates for each aisle; this pattern is suitable for a small system used by a dealer.
- Pattern II - All aisles are connected by a loop conveyor line, thereby integrating receiving and delivery. This is the most typical pattern for large-scale systems.
- Pattern III - This is a revision of Pattern II. In order to increase the throughput capability, the loop line is combined with direct transit lines.

This paper introduces an automated warehouse system for cars, taking the example of Nissan Motor Oppama Wharf, which was completed and delivered in July 1984 and boasts the latest and largest facilities.

System outline

The whole system (Fig. 3) consists of storage/retrieval cranes (IHI Auto Stack Crane) operating in the high-rise storage structure, receiving/delivery stations for taking in/out cars, and CARTRAC loop line (a typical car-transfer system) connecting the cranes and stations (Figs. 4–8). All are operated under the on-line control of an exclusive process control computer.

The specifications of these facilities are as follows (note that future expansion is taken into consideration in designing these specifications since the expansion of the current capacity to twice as much is scheduled for the future):

- Objectives: all types of passenger cars and some commercial vehicles
  - overall length 4825mm maximum
  - overall width 1870mm maximum
  - overall height 1555mm maximum
  - weight 1535kg maximum
- Storage capacity: 10,488 units (20,976 units)
- Throughput capacity:
  - Receiving: average 1200 units/day (2000 units/day)
    maximum 1300 units/day (2200 units/day)

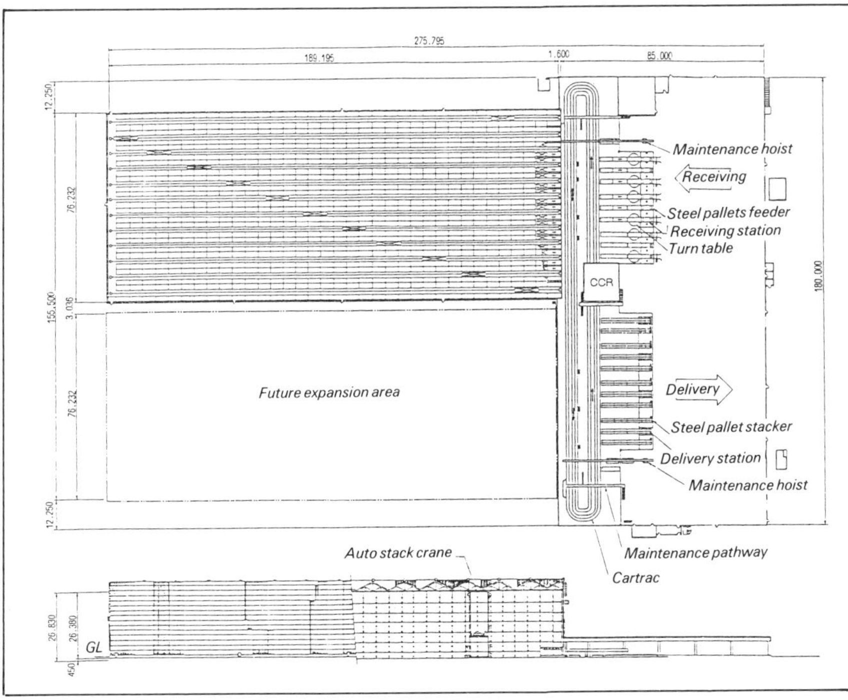

Fig. 3 General layout of warehouse

- Delivery: average 1200 units/day (2000 units/day)
  maximum 1700 units/day (2500 units/day)

- Auto stack crane:
  - Number of cranes - 12 sets (24 sets)
  - Speed: travelling - 0–150m/min
    hoisting - 30/7.5m/min
    forking - 30/10m/min

- CARTRAC:
  - Number of cars - 30 sets (53 sets)
  - Speed: travelling - 60m/min
    loading/unloading - 22m/min

- Conveyor:
  - Receiving line - 4 lines (6 lines)
  - Delivery line - 5 lines (7 lines)
  - P & D line - 12 lines each (24 lines each)

- Process computer:
  - HIDIC-80 (128kW)
  - CPU and peripherals

Note that the numerals in parentheses are those after expansion, and while the system operates for 24 hours a day on two shifts, the throughput is shown based on operating for 420 minutes/day.

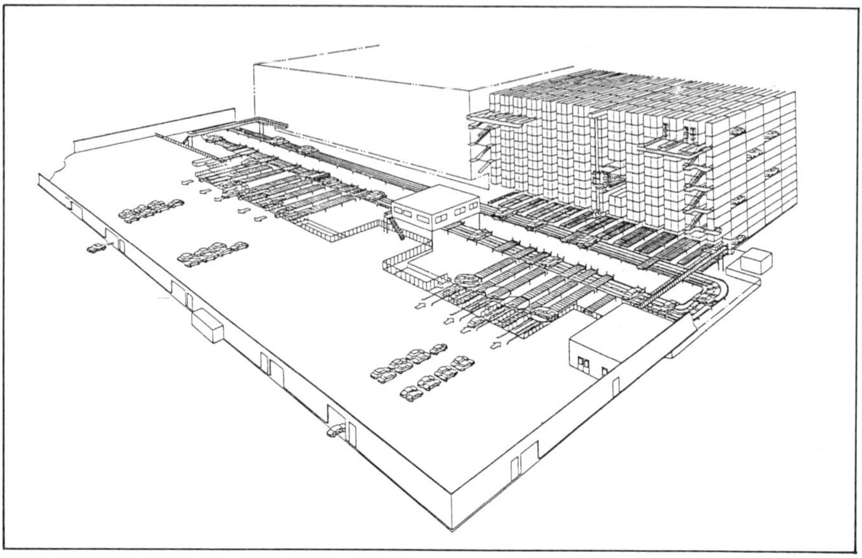

Fig. 4 Schematic of receiving and delivery area

Operating procedures

This automated car warehouse is a subsystem of Nissan Motor's total export system, and the process computer for the automated warehouse control communicates directly with the mainframe computer controlling overall data processing. Accordingly, data related to export procedures, shipping plans and accounting controls are processed by the host computer, leaving car location controls and automatic controls of all equipment as the major tasks of this process computer. The procedures from receiving to shipping based on the car flow are as follows.

Fig. 5 External appearance of warehouse building

Receiving and storage system

After leaving the assembly lines of each plant, the cars are taken to the wharf by trailer trucks. All information on the cars being transported is controlled by the host computer and the location of each car can be known throughout the route.

Cars unloaded from the trailer trucks are driven to the receiving space of the automated warehouse and loaded by the driver onto a steel pallet supplied automatically to a receiving station. At this time, the punched card attached to each car is read by the machine-side card reader, and all records are registered in the process computer. After this, each car is stored at the optimum location in the storage structure by the conveyor, CARTRAC and Auto Stack Crane under the control of the process computer.

Since the receiving station functions as the contact point between operators and automated equipment, gates with various indicators and external sensors are installed for the safety of operators.

Retrieving and shipping

As soon as the host computer completes procedures necessary for shipping, the related data are transmitted to the process computer as orders for shipment. Upon receipt of these orders, the process computer

Fig. 6 Storage racks and Auto Stack Crane

assembles the data necessary for shipping and prepares a retrieved booking file for the start of shipping work. The shipping supervisor inputs the key-code according to the preparation situation at the berth and checks the invoice through the CRT terminal. The required cars are carried to the delivery station in order by the Auto Stack Crane and CARTRAC.

When each car arrives at the delivery station, a shipping-mark tag is output by the printer installed at each delivery station. The operator checks the tag for the necessary items, pastes the tag on each car and unloads the car from the pallet. Each car is carried from the delivery area to a specified berth and shipped.

In addition to the normal delivery for shipping, exceptional retrieval for quality assurance inspection or customs inspection is possible with this automated warehouse.

The delivery station is also provided with various equipment for the safety of the operators in the same way as the receiving station.

Features

The most important aspect of designing this automated warehouse was how to realise a throughput able to cope with the speed with which export cars are shipped. Accordingly, the system balance was checked

Fig. 7 Receiving station at 'ready' condition

Fig. 8 CARTRAC at radius-turn

under various working conditions based on computer simulation of the whole system, in addition to theoretical analysis, before determining the specifications of all facilities. The following improvements were made to the actual design by feeding back these analytical results:

- The capacity of CARTRAC was increased by 50% or more by adopting the optimum layout and improved algorithm.
- The capacity of the Auto Stack Crane was increased by 20% or more, increasing the travelling speed as a result of adopting an inverter control method and by revising the allocation logic.
- Optimisation of the number of stations and number of conveyor storage locations.

The scheduled capacities were confirmed during the performance test conducted at the time of overall trial operation and, as a whole, this automated warehouse has a very high performance.

Minimised down-time

Minimising down-time is an important theme for any facility. This is especially true with this automated warehouse which directly relates with strictly timed shipping work; this has been reviewed very carefully. In planning this automated warehouse, all troubles were analysed by cause based on the operation reports of existing automated warehouses. These were evaluated quantitatively and 17 improvement items were selected, some of which are outlined below:

- Adoption of tilting mechanism for the receiving station.
- Reviews on the system tolerances and operating clearance.
- Easy maintenance of conveyor units by adopting simplified modular-component design.

As a result of taking these measures, the intended targets of reducing the frequency of malfunctions to one quarter of conventional systems and keeping MTTR within 15 minutes after a problem has been reported, have been sufficiently accomplished.

Storage structure with superior resistance to earthquakes

As the scale of facilities becomes larger, behaviour in earthquakes becomes more complicated and the influence of disasters becomes greater. Therefore, a dynamic seismic analysis was constructed with a computer in addition to normal state analysis, in designing the storage structure. The following was learnt from these analyses:

- The fundamental period is about 0.5s.
- The response of acceleration at the structure top is about five times as much as the ground acceleration.
- Seismic factor specified in 'The Building Standards Law of Japan' is equivalent to nearly 1.5 times the computer-analysed force.

The suitability of the results of dynamic analysis were verified by conducting a series of vibration tests using a full-scale partial model.

All members were checked for safety, based on these studies and test results, giving the structure a superior resistance to earthquakes. Precautionary measures were taken for this, for example, installing stoppers for stored items.

Advanced control system using microelectronics

Up to now, a centralised control method has been adopted in which signals from sensors, indicators and other miscellaneous devices are fed to a centralised console in the CCR (computer control room) through field control panels and multi-core cables (Fig. 9). The console is connected with a process computer through a process I/O relay unit.

However, the control system of this automated warehouse consists of a type of local area network (LAN) in which all signals are integrated in each block and converted to serial signals by the interface for two-way communications with the computer and console panel in the CCR.

This system has three sets of high performance sequence controllers linked with the process computer and dedicated high-speed channels, realising a hierarchical and decentralised control system. Therefore, the whole control logic is incorporated in software, which not only increases the flexibility to system changes but also substantially reduces the number of process I/O and wiring, making the whole system simple and easy to maintain.

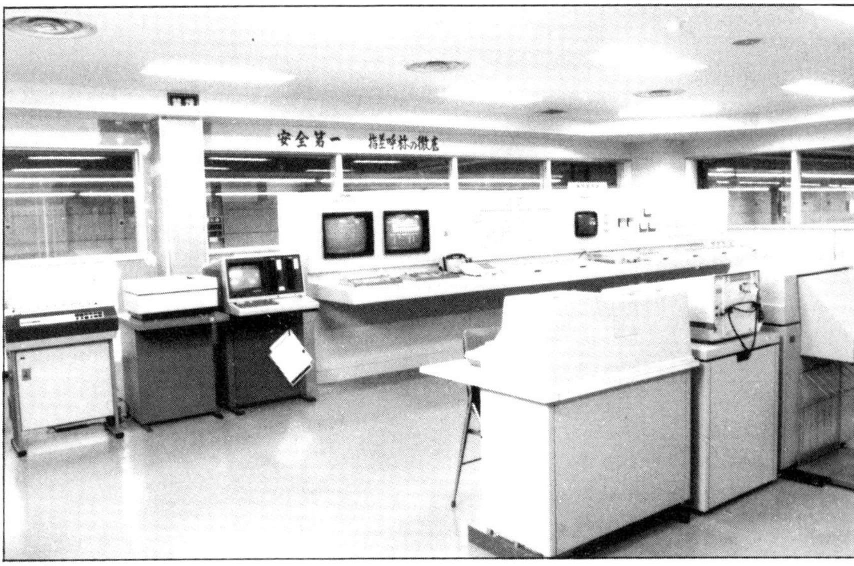

Fig. 9 Computer control room (CCR)

Extended operation functions

IHI has attempted to extend the operating support functions as much as possible in order to cope with the user's needs of labour saving, energy saving, and higher maintainability, all of which have become important recently. The major extended functions are:

- Automatic routine inspection system. Each crane automatically checks according to specified logic by remote operation from the CCR and the results are displayed on the operator's console.
- On-line data gathering of actual operations. Data, such as operating time, number of cycles and alarm contents from each equipment, are stored in real-time by the computer database, and the data can be retrieved as needed in any required format.
- Automatic safety guard against earthquakes. Earthquakes are detected by field sensors, and the operation is automatically stopped in three stopping modes according to the seismic magnitude, thereby protecting all equipment from possible damage.
- Intelligent ventilation system. Electricity consumption can be saved since the forced ventilation is automatically started and stopped as a sensor judges the weather conditions.
- Integrated graphic console. All functions related to monitoring the operation of all equipment are integrated on the operator's console, and the information is displayed in graphic form on a colour CRT.

In the past, a desire to incorporate all these functions increased the cost, and almost all were abandoned as mere ideas. However, IHI attempted to establish a software-orientated control system with this project, as well as realise all these ideas in the actual system.

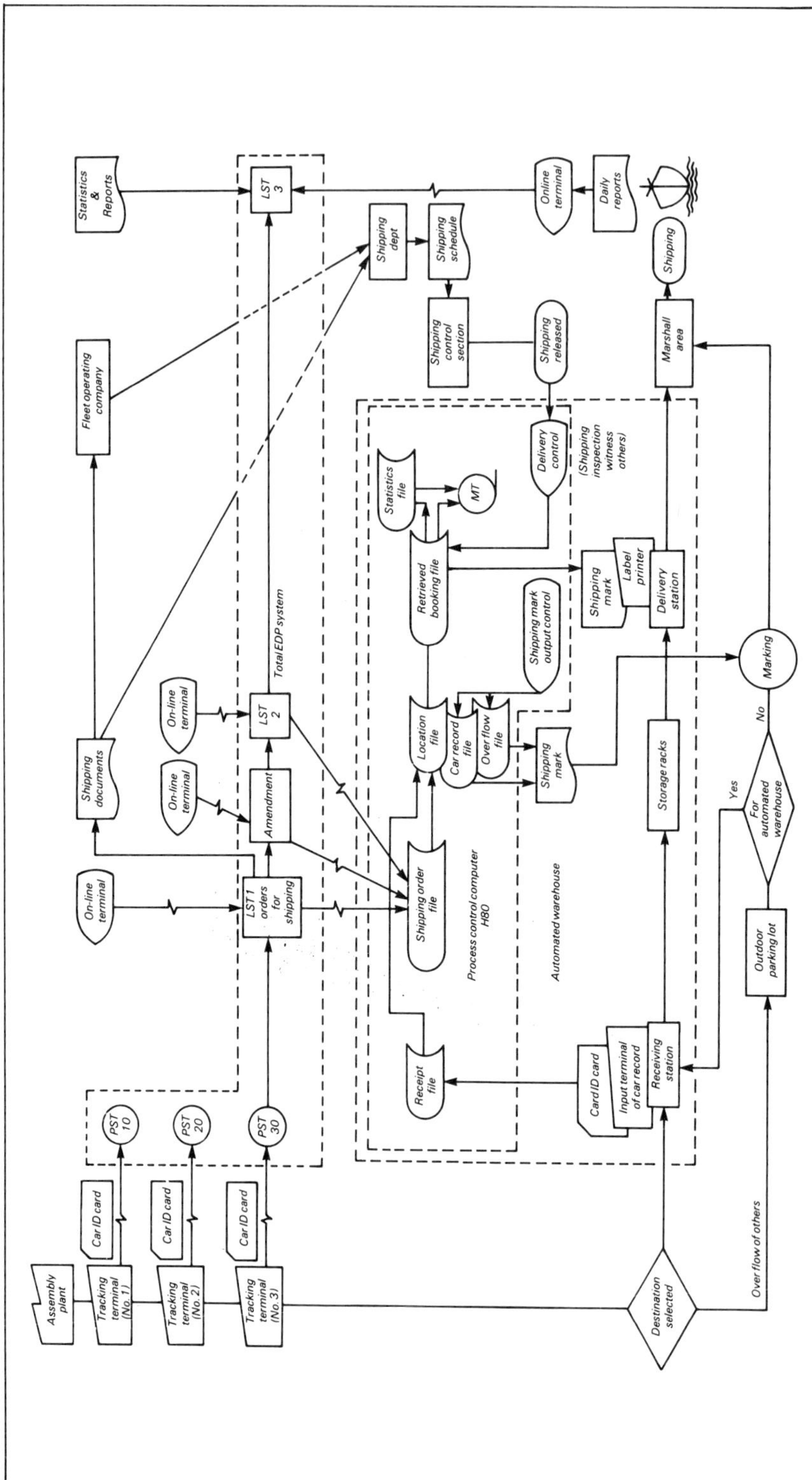

Fig. 10 Nissan's total export system and procedures

Concluding remarks

Car storage using automated warehouses has many advantages over conventional systems: space efficiency, less labour, quality maintenance, and safety. All these benefits have been confirmed by this case study which also indicated the following possible advantages from the viewpoint of handling cars for export:

- As customs clearance has been simplified and the lead time is shortened, temporary storage before shipping can be reduced.
- Since a large number of cars can be gathered on one wharf, the loading efficiency of cars on large ocean-going car ferries can be increased.
- Since the shifting and marshalling of cars by drivers in the storage yards can be minimised substantially, damage to cars before shipment can be reduced.

As described, this automated warehouse, which has been constructed as a part of IHI's efforts to rationalise the distribution of export cars (Fig. 10), has a high performance and a variety of features because of the latest technologies applied to it. It is believed that this automated warehouse will fully demonstrate its functions and answer the many expectations of the export field in which requests for more speedy and better services are increasing continuously.

Acknowledgements

The author would like to express his sincere gratitude to all related persons in Nissan Motor Co. Ltd for their kind cooperation and valuable advice given through the whole process of the warehouse construction and operation.

SIMULATION IN ADVANCED MANUFACTURING TECHNOLOGY

I. Langley
Istel Ltd, UK

This paper describes how simulation was used as the central vehicle for progressing a design study for a flexible manufacturing system. The company concerned is a well established supplier of formed metal components to the medium and light manufacturing industry. The forming process imposes heavy wear and tear on the forming dies. Consequently there is a regular demand for both new and replacement dies to maintain production.

The manufacture of forming dies currently follows normal tool-room practice, with associated long lead-times and high investment in dies at various stages of completion.

In order to improve its response to changing customer requirements, the company decided to invest in a totally new facility which would exploit the latest developments in manufacturing technology. The company had the foresight to realise that this entailed considerably more than advanced metal removal techniques.

When Istel was first involved, a significant amount of work had been done. The company had completed an analysis of the total range of components; these fell into two main categories, cuboidal and turned. Further subdivisions took into account such factors as whether the die was hard-treated or not, and which processes were required. The basic accesses were: soft-stage machining, heat treatment, skimming and electro-discharge machining.

At this stage the company had not decided on the precise configuration of machines nor on the allocation of processes to machine types. However, detailed estimates of individual process times were available for all stages of manufacture for each class of die.

The problems the company wished to address for a given volume of requirements were:

- How many machines of each type are required?
- What level of inter-process travel was generated?

- What level in work-in-progress (WIP) occurred?
- How sensitive was the system to machine breakdown?
- How was the system performance affected by part programming trials?/

It was agreed that the most reliable answers could be achieved from a simulation study.

The first simulation model

The die shop was modelled as a series of machine cells, each cell containing one or more identical machines. Each machine cell had an associated work-in-progress queue where parts would wait until a machine in the cell became available.

Work was generated on a daily basis. The parts were moved from one machine cell to another, joining a WIP queue if necessary. The route through the shop and the process times at each stage could be specified for each part.

Material handling was not modelled in detail. A delay factor dependent on the distance between machine cells was built in to represent travel. The simulation did not model AGV movement or availability.

The layout of the die shop and the parts in the shop were displayed on a colour visual display unit (VDU). It was, therefore, possible to monitor the progress of individual parts through the shop and to observe the build-up of WIP queues.

In addition to the visual display the simulation collected statistics on lead times, work-in-progress queues, machine cell utilisations and daily throughput.

Many parts undergo heat treatment out of the shop. The simulation monitored the number of parts being treated, but did not model the process in detail. A constant turnaround time in heat treatment was used.

It should be borne in mind that, for the purposes of this first stage model, all machines were assumed to be 100% efficient. It was, however, possible to interact with the model to cause machines to be unavailable, thus representing down-time.

The simulation generated new work on a daily basis using the specified annual production quantities. The model incorporated a means by which a fluctuation in the work issued each day could be produced. There was also the facility to allocate to a certain proportion of parts providing status. These parts would take twice the normal time to be processed on NC machines, and would thus represent the effort spent in testing new part-programs.

Model configuration and operating rules

The first die-shop configuration to be investigated contained the following cells: milling, milling inspection, turning operation 1, turning

Table 1 Number of machines required in each cell to achieve required output

Machine cell	*Number of machines in cell*
Milling	*2*
Milling inspection	*1*
Pre-heat turning	*2*
Post-heat turning	*1*
Sparking (small dies)	*4*
Sparking (large dies)	*7*
Lettering	*1*

operation 2, sparking (small dies), sparking (large dies), and lettering. The first turning cell performs operations on any dies which require turning before undergoing heat treatment. The second turning cell processes any dies which require turning after heat treatment.

The electro-discharge machines (EDM) or sparking machines were divided into two cells. All dies of less than a certain length are processed by the 'small' sparking cell. All others are routed through the 'large' sparking cells.

In the event of a die undergoing two consecutive operations in the same machine cell, e.g. milling, priority was given to jobs which were partially completed. In all other cases, jobs were selected on a first-in first-out basis.

Experimentation procedure

The simulation model was used to investigate the performance of several die-shop proposals. This section describes the results for the final configuration agreed.

The model simulated ten weeks of die-shop operation. The results described in the following section are based on the statistics recorded during the runs.

After some experimentation it was concluded that the required output could be achieved, under a three-shift working pattern, if the die shop operated with a certain number of machines in each cell (Table 1).

There was congestion by the first operation (either milling or first turning) at the beginning of the day after the work for that day was issued.

The work-in-progress queues recorded at the other machine cells were small in comparison, as earlier processes smoothed out the flow parts to later operations. The average lead-time, across all parts, was 52 hours.

Whilst the transport mechanism was not modelled in detail, the results suggest that one AGV would be sufficient to handle the transfer of parts between machine cells.

After some deliberations by the company it was decided to proceed with a detached specification of a manufacturing system based upon the following configuration:

- One central material store.
- Three NC lathes.
- One NC machining centre.
- Three large electro-discharge machines.
- Two small electro-discharge machines.
- One inspection machine.
- One AGV system with two or three vehicles.

This specification had to include the control and communication systems which would link the machine tools and the mechanical handling equipment.

It was agreed that a detached simulation was the only way that a satisfactory degree of integration could be achieved at this stage. In this case, the questions to be addressed were:

- AGVs – How many ? What track layout? What control/priority rules?
- Machines – Where to position? What priority? Determine workplace layout and buffer capacity.
- Stores – Manning requirements?
- Working patterns – Explore end-of-shift effects. Explore start-up conditions.
- Shop control – Establish work issue rules.
- Overall performance – Determine lead-times and WIP.

A number of shop and track layout and operating patterns were tried before the final arrangement was agreed. In this layout the machines were linked by a one-way AGV track. These three turning machines and the five electro-discharge machines were each served by three 'pick up and drop' (P&D) stations, while the single machining centre was served by six P&D points. Each point was served by a spur off the AGV track. The AGV supplied components from a central store to each of the machines, and for the electro-discharge machines only, transported the roughing and finishing electrodes.

To improve the response of the transport systems to the EDMs, an intermediate buffer store was established in the centre of the shop adjacent to the EDM section. This buffer store was used to hold both tooling and parts. Arrangements were that the machining centre would need to run, manned for three shifts, whereas the lathes would run for two shifts only. The EDMs would be manned for two shifts, and because of their long cycle times (typically four hours and more) would run unmanned into the end of cycle in the third shift. The AGV system would run for two shifts only.

The first and most difficult exercise with the simulation was to derive a set of navigation rules. A prime requirement was to minimise travel empty time; so wherever practical, the AGV took an unprocessed part to a machine when it collected a completed one.

The navigation rules had to resolve a particular conflict. On the one hand it was necessary to prevent collisions and ensure that only feasible tasks were assigned; on the other hand it was possible to be over-cautious, and create situations where each AGV would be waiting for another move. These rules were further complicated by the complex sequence of component and tool changes which occurred on the electro-discharge machines.

The model displayed all the machines and the AGV track on the screen in approximately their true position. The AGVs were shown moving around the track collecting and delivering stock and tooling between the store and the machines.

In addition to this representation of the functioning of the shop, the model monitored the following: machine states, AGV states, component lead-times (by route), despatches from store, receipts into store, WIP (by process), AGV visit frequency, time in shop (by process), and work outstanding at end of each shift (by process).

Once an efficient set of rules had been established, a number of points emerged. Firstly , that the shop efficiency was sensitive to the order in which work was moved. With a limited buffer capacity, the machining centre processed the daily issue of parts in order of ascending process time. Reserving the long cycle work for the third shift maximised the utilisation of this machine.

One consequence was that the following morning it took up to one hour to set all the machines running, as the AGVs could not cope. By adopting the reserve policy for the lathes, such as issuing long cycle work first, it was found that the shop was fully operational within 30 minutes.

This second stage model was driven by data reflecting the true working pattern, and the resulting response of the model was used in deriving a scheduling method which smoothed the work load while maximising the use of the installer capacity.

Experiments with the most recent design showed that lead times of one week or less were achieved consistently across the whole parts range, and that the economical arguments for the system are most convincing.

Concluding remarks

Neither of the two simulation models solved any problems. However they were invaluable catalysts to the system designers, for they presented them with advance knowledge of the operational conditions which would arise. The designers were thus able to concentrate their efforts on achieving a pragmatic and efficient system.

A GUIDED VEHICLE SYSTEM FOR THE ASSEMBLY OF ENGINES ON THE MOVE

I. Langley
Istel Ltd, UK
R. Meek
CFC UK Ltd, UK
and
N. Phillips-Parham
Land Rover Ltd, UK

Land Rover, a world leader in four-wheel drive vehicles, has been modernising its manufacturing facilities onto one site at Solihull, UK. A feature of the new facilities is a guided vehicle system for the assembly of engines. This paper describes that system and how simulation has been used for the successful introduction of the system.

What is claimed to be the first installation in the UK to assemble engines on-the-move on automated guided vehicle workstations is in operation at Land Rover.

Land Rover is a world leader in the four-wheel drive vehicle market. And in order to remain competitive the company has invested heavily in modernising and rationalising manufacturing facilities onto a single site in the UK.

In order to achieve the greatest possible benefits from this modernisation and rationalisation programme, Land Rover made extensive use of the SEE WHY computer simulation package from Istel The simulation has been used in the successful introduction of the automated engine assembly system.

Simulation

The facility

The facility considered in this exercise was new, designed by contractors to use automated guided vehicles (AGVs). The process consists of kits of parts being loaded onto the AGV which then follow a series of loops or workstations. The operator mounts the vehicle, carries out some engine build operations, alights from the moving vehicle, returns to the start of the loop and repeats the operation on the next AGV.

Following the initial assembly operation, the AGV computer chooses one of two paths, depending on engine derivitives. Basic engines going direct to unload whereas more complicated engines proceed via a number of special build cells for additional work before being unloaded.

All engines are subsequently loaded onto an overhead conveyor that transports them to test cells. After testing, engines are loaded onto a separate AGV circuit where final assembly operations are completed.

Why simulate?

With any new facility there are a number of queries. Land Rover divided these into two areas:

- Facility – How many AGVs are required? Is there sufficient allowance for AGV battery charging? Will the software control rules specified give the required performance? Are there any serious bottlenecks?
- Operational – What control should be placed in loading of AGVs? What should be done in the event of an AGV breakdown? What are the implications of a change in engine mix and/or volumes?/

Clearly it was essential that facility queries were answered as soon as possible, before irreversible commitments on expenditure were given to the contractors. The operational queries although important were less urgent than the facility questions.

In order to find solutions to these questions rapidly and with the required accuracy, visual interactive simulation using SEE WHY package was chosen. This technique enabled both facility and operational queries to be answered, and provided an effective way of demonstrating solutions to both Land Rover management and contractors. In addition, the availability of an interactive simulation would enable the model to be used to evaluate future possible modifications of the facility.

The model

Following a short period of data collection, a detailed simulation model of the facility was designed and developed using the SEE WHY package. The model allowed engineers to interactively change many parameters, e.g. number of AGVs, cycle times of processes, speed of AGVs, engine mix required, duration and likelihood of AGV breakdowns, etc.

The engineers were able to verify that the model was a true representation of the facility by watching the dynamic colour schematic of the facility on the screen. Following verification, the SEE WHY model was used to answer the facility queries. Demonstrations of the model experimental results were regularly carried out with the contractors and Land Rover management until satisfactory solutions had been agreed. Following this, the operational queries were investigated and the solution implemented.

Results

Use of the model allowed significant changes to be made to the facility, safely. These included a reduction in the planned number of AGVs even with an engine programme with the highest work content.

In addition, changes were made to the planned control at the load point. This was necessary because when an AGV broke down it caused other AGVs to fill free positions in the system which led to a shortage at the load point. The control was such that a shortage at the load point forced AGVs to be released from battery charging, which in turn could lead to some of them stopping because of lack of charge.

The exercise proved computer simulation to be of considerable benefit in the design and implementation of new automated manufacturing facilities. Visual interactive simulation allied to a committed team of consultants, engineers, management and contractors has the potential to reduce drastically 'teething problems' on implementation of new facilities and to determine the precise facility requirements prior to final commitment of finance.

Guided vehicle system

The flexible system for the assembly of V8 engines at Land Rover's Solihull plant has been designed and installed by CFC UK Ltd. The system provides Land Rover with enhanced planning and production control in engine manufacture (Fig. 1).

Production of 3.5 litre V8 engines for Land Rover and Range Rover has been transferred from the company's works at Acocks Green in Birmingham, to its main plant in Solihull, sharing the same production area as the four-cylinder engines used in Land Rovers.

The move is part of a major rationalisation programme aimed at improving efficiency and increasing productivity in which all the company satellite factories in Birmingham have been closed and component manufacturing transferred to the main Solihull site.

Having decided to move engine production, Land Rover was faced first of all with the problem of whether to move existing plant from Acocks Green to Solihull or invest in new equipment. After selecting the latter course, it had to decide which new assembly system to adopt.

Conventional engine assembly lines using floor-level conveyor systems have a number of disadvantages when compared with Land Rover management's requirements for the new installation. Such conveyors can require a long, straight floor area and expensive underfloor works to accommodate the return of the conveyor, and their lack of flexibility means costly modifications if a new model is introduced.

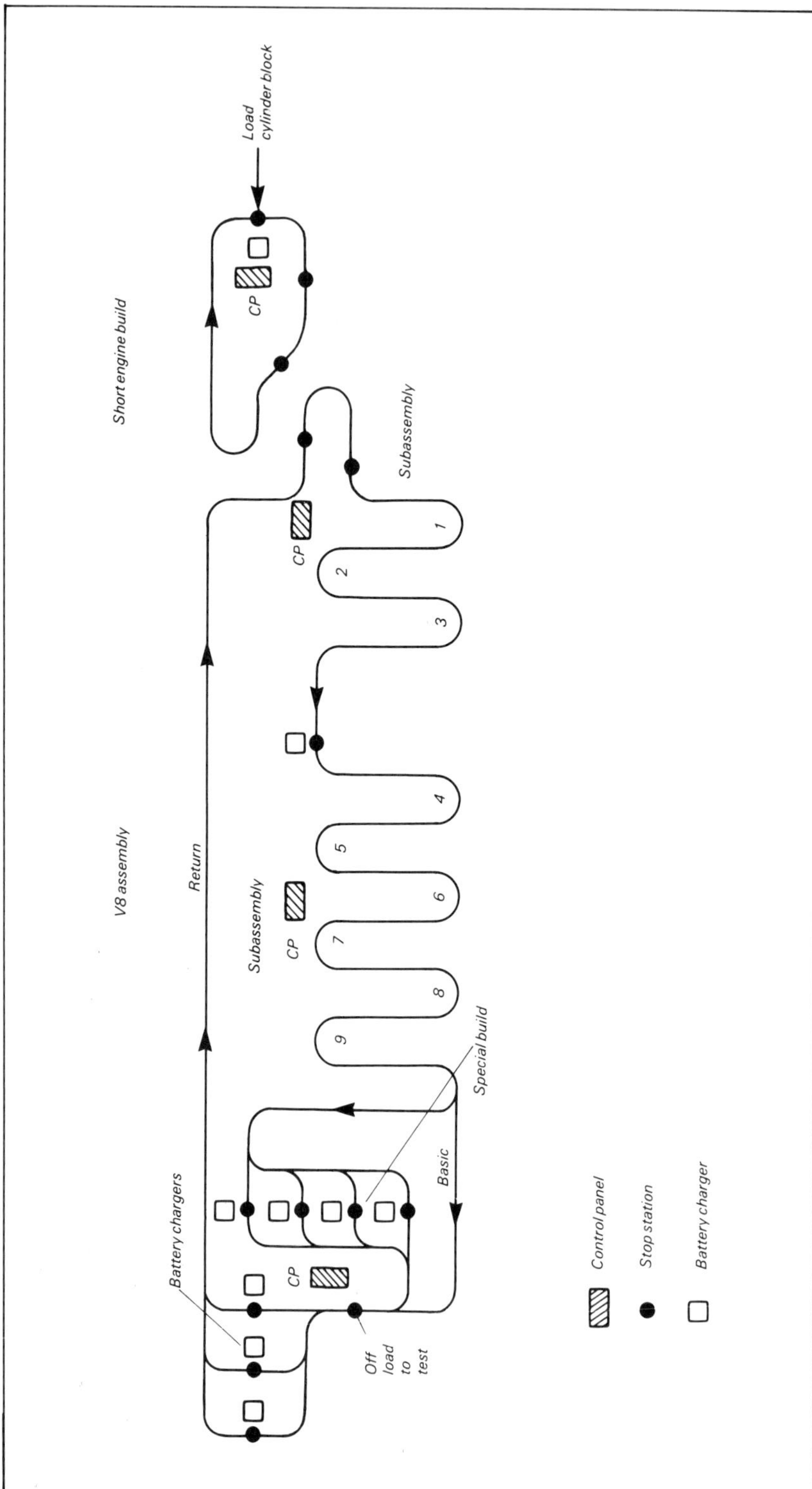

Fig. 1 Layout of V8 assembly and short engine build

The prime requirement of the system

A prime requirement of the production managers at Solihull was that the new facility should occupy a total floor area of 80 × 17m, and that it should be flexible in operation. After an investigation of the latest assembly methods and techniques in use throughout Europe, the use of automatic guided vehicles was selected as the method most likely to give the required productivity, versatility and reliability. CFC UK was chosen to design and install the system in close collaboration with Land Rover.

In addition to flexibility, the AGV system is said to provide an improved working environment, can be modified or extended at low cost, is easy to maintain, and offers a high level of safety.

A team charged with planning the system included production engineers, supervisors and operators, industrial facilities engineers, and CFC representatives. By encouraging all those connected with engine assembly to comment on various proposals, commitment helped achieve the objectives of improved quality and productivity, and reduced inventory for work-in-progress.

Three independent AGV systems have been installed to cater for initial V8 short engine build, main V8 engine assembly, and final dressing of both V8 and four-cylinder engines. Altogether, 70 AGVs are employed, all of which are interchangeable between the system according to work load.

Additional AGVs can be added at any time to increase productivity, while any AGV that needs to be taken off the circuit can be withdrawn without interfering with production, thus allowing much more accurate management planning.

System flexibility can cater for many engine derivatives, including carburettor engines, petrol injected engines, air-conditioned vehicles and power steering variants. Altogether there are seven base models with 49 variations of the V8 engine and ten base models with 157 variations of the four-cylinder engine.

V8 engine manufacture

Cylinder blocks for V8 engines are machined in another area of the factory and then delivered to the load station to be mounted in a fixture on an AGV. All fixtures used on the AGVs in engine assembly are capable of both rotation and tilt to enable the operator at each station to adjust them to the most convenient position for each assembly operation.

In this section, the crankshaft, camshafts, clutch, flywheel and starter motor are fitted to the block at five workstations. Here the AGV is stationary during assembly. The block is then conveyed to the next station.

At the unload station, there is a small buffer store to ensure there are always blocks waiting to be loaded into the main assembly section. The

AGV returns to the load station where it receives a short battery charge before restarting the cycle. The loop is 45m long.

In the main V8 engine assembly section, all work is carried out while the AGV is on the move. The track is laid in a series of loops rather than a straight line, allowing 90m of assembly track to be accommodated in a floor length of only 31.5m.

Each operator's assembly task is completed while the AGV travels round a 10m loop, leaving the operator with only a few metres to walk to board the next AGV. This is said to reduce fatigue and possible damage to the assembly, and increase efficiency and productivity.

At the load station, a block from the short engine build section is again mounted on a fixture. Engines for basic and special build are loaded onto AGVs at random; an operator identifies the engine to the control system by pressing one of two pushbuttons in accordance with the engine build card. A light in the pushbutton gives a permanent display of 'basic' or 'special' throughout the assembly cycle.

Should the operator make an incorrect selection, it can be changed provided the start button has not been activated. The AGV cannot move until a selection has been made. After the start button has been pushed the AGV proceeds to the first workstation.

Throughout the assembly track, there are several 'stop' positions where an AGV will come to rest if there is a vehicle already on the next section. It restarts automatically when the track is clear. Selected stop positions have battery charging points where a short boost charge is given automatically.

The two types of engine are routed by the AGVs following the selected frequency from one of three wires of the four-core cable buried in the floor. The fourth wire provides two-way communication between the AGV microcomputer and the control computer.

During assembly, the AGV travels at a speed that can be varied at the control station to any one of 32 values between 0.75 and 3m/min according to production requirements. At other times it travels between stations at the higher speed of 35m/min.

Leaving the load position, the AGV travels to the first kit-marshalling point where components and subassemblies are loaded into trays on the AGV. This reduces lineside stocking and creates more floor space.

Parts are assembled during the first three loops of the track after which the AGV stops at a second kit-marshalling point, which also acts as a buffer. Small value items such as nuts, bolts, washers, etc. are held in special storage units adjacent to the track from which the operator can help himself as the AGV passes.

After travelling through another six assembly loops, the basic engines are complete and routed to an unload station; 'specials' are routed to a third kit-marshalling point where components are loaded according to the specification on the engine build card. There are four stationary special assembly stations; the AGV travels automatically to the first one to become free.

After fitting the additional units, an AGV is released to the unload station where engines are lifted to a CFC Railmatic monorail overhead conveyor, which carries them to the test facility.

Each AGV is given a 21 minute charge before returning to the load position; however, this can be over-ridden should production requirements so demand it.

An Allen Bradley 2/20 PLC records the time an AGV takes to travel between check points throughout the assembly track. If a vehicle is overdue a signal is transmitted to the controller to give an alarm and indicate where a particular AGV is located. Over one preselected section of the track, the actual speed of an AGV is measured and compared with the programmed value.

Full information about the state of the assembly section and the movement of the AGVs is displayed in colour on a graphics monitor in the supervisor's office, together with production information such as target output, speed programmed and actual speed. The microcomputer on board the AGVs and the host computer in the supervisor's office provide complete two-way communication that allows control of production rates.

The supervisor can blank out certain sections of the track depending on production requirements. For example, depending on the number of specials in a daily target, any number of the four special assembly stations can be made inoperative.

Should there be a computer fault in the office, semi-automatic control of assembly can be taken over at the main electrical cabinet alongside the track. If automatic communication breaks down, the AGVs can be controlled manually from this panel.

Final dress of four- and eight-cylinder engines

After passing through the test facility, both four-cylinder and V8 engines are delivered to the final dress section by the overhead monorail conveyor. The AGV fixture is identical for both types of engine, which means the AGV can arrive at either load station at random (Fig. 2).

Control and operation in this section are similar to those in the main V8 assembly; the exception is that AGVs are stationary during assembly operations. AGVs are routed through the section and engines equipped in accordance with the travelling build card.

When engines are complete, they are delivered to the unload station and put on to a normal conveyor for transfer to other sections and despatch. Managment control provides for the switching out of either the complete V8 or four-cylinder dress line according to production requirements.

In this section, a number of inter-station routes cross main gangways and the control reduces the 35m/min fast travelling speed to 5m/min in these areas, ensuring that only one AGV is on the gangway at any time.

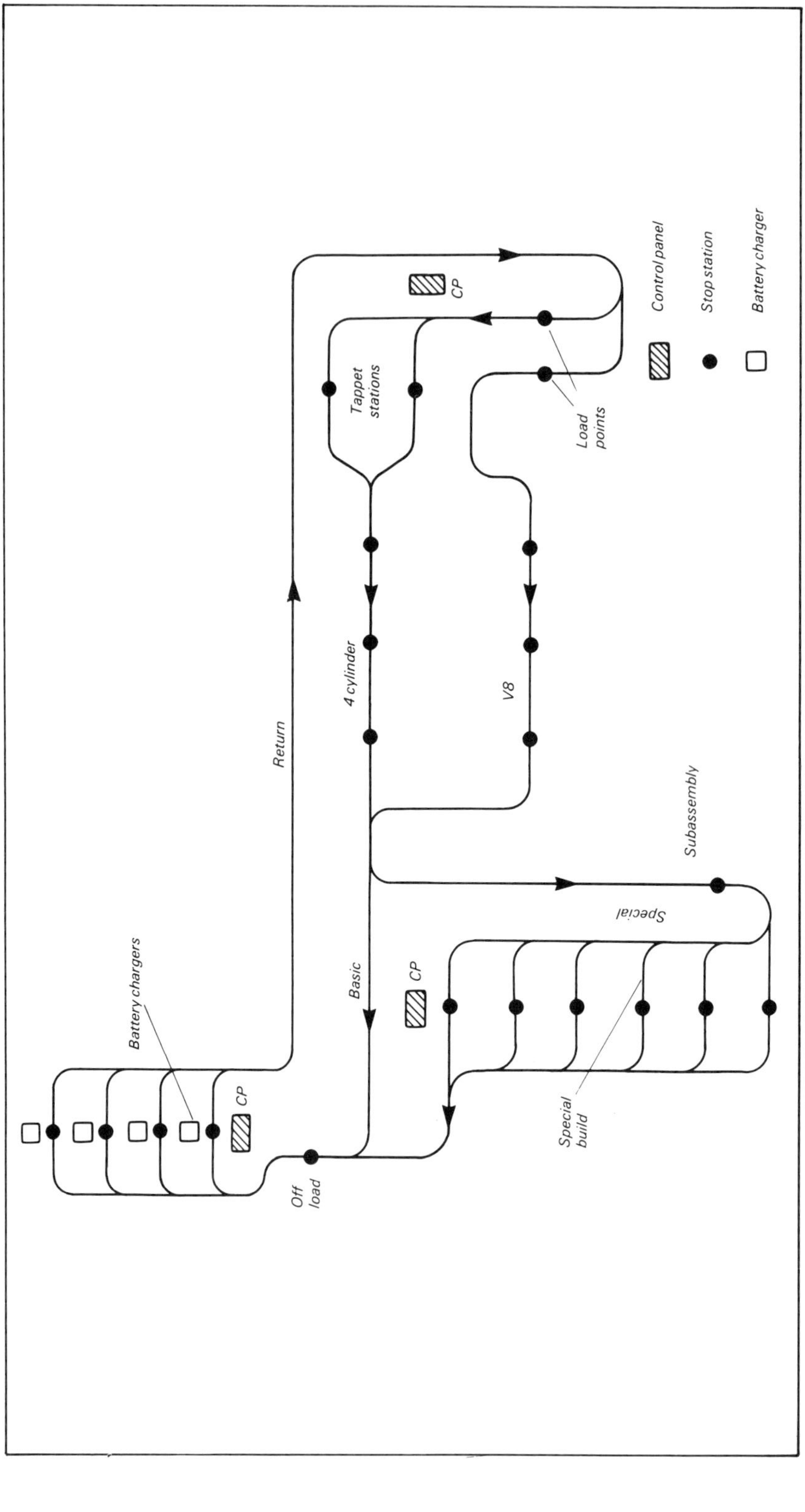

Fig. 2 Layout of final dress

The automatic battery charge after the unload station is of 10 minutes duration.

The assembly-type AGV used at Land Rover is fitted with anti-skid footboards set at 105mm above floor level on three sides. Several different sizes of AGV are available according to the applications. Those at Land Rover being 1340mm wide × 730mm high × 1800mm long.

Each AGV has two operating panels, both of which have a start push button and an emergency stop button. In addition, the right-hand panel has a meter to indicate the state of the battery, including a green light to show a fully-charged condition and a red light to show when capacity is below minimum. A multi-pin plug provides the connection to the manual control box.

The left-hand panel houses the basic/special select buttons, on/off key switch, reset key switch, and a light indicating that guidance sensors are correctly located above the operating frequency wire in the floor.

The reset switch is the only way in the assembly line an AGV program can be changed once it has left the loading station, and is used only by an authorised supervisor. For example, any problem arising on a basic engine that could cause a hold-up in production can be re-routed to be dealt with in a special station.

The need for safety devices

Two major safety devices are included. A front-mounted ultrasonic sensor beam detects any object in the path of the vehicle, causing it to stop, When AGVs are in a stationary queue, for example at loading, unloading or kit-marshalling stations, the sensor beam creates space between them to allow operators to cross the line freely. In addition, a sensitive front bumper causes the AGV to stop within the travel length of the bumper if it contacts any obstacle.

If an AGV needs to be taken from the guide-path, a hand-operated control can be plugged in and the AGV driven off the track. Should the fault be in the drive system, a handle allows the drive wheel to be lifted off the floor and the AGV easily pushed off the track. Again, production can be maintained.

Electrical control equipment and the drive unit are readily accessible for maintenance. Automatic diagnostics are built into the system.

All Land Rover employees involved with the move from Acocks Green to Solihull were kept informed of company plans. CFC UK installed a short track at Acocks Green to allow 33 operators to be trained in the new methods over a period of 11 weeks prior to the move.

The order was placed with CKC UK Ltd in December 1984 and the four-cylinder engine dress section was operational on the 2nd of September 1985 followed by the V8 short engine build and main assembly lines on the 30th of September 1985 and the V8 engine dress section in November 1985. This three-phased move ensured continuity of production and avoided the need to stockpile engines. All the deadlines were met and the planning and training ensured a smooth transition.

On the V8 short build and main assembly sections, productions on a one-shift basis ceased at Acocks Green immediately prior to a one-week works holiday and started up at Solihull on day and night shift on the return.

It is claimed that production reached 100% of line rate within four weeks – tribute to the close cooperation between CFC and Land Rover engineers throughout the planning and installation periods.

Land Rover will carry out planned and breakdown maintenance on the system, and a small track has been laid by CFC in the maintenance section where all the functions of AGVs can be tested. Personnel have been on a comprehensive maintenance training programme at CFC's headquarters in France.

The system allows production volumes to be varied, or different models to be produced. Another effect of the new working methods and longer work cycles is that work traditionally associated with repetitive assembly is less monotonous leading to an increase in overall efficiency. Further job satisfaction has resulted from making operators responsible for the quality of their work.

For enhanced management control, the information presented on the colour monitor in the supervisor's office could be relayed to the production director and other management personnel.

Although the new systems have only been operating for a short time, Land Rover is able to report significant benefits, including higher quality through improved working methods, 6% reduction in hours/engine, and a lower work-in-progress inventory from over 400 engines at Acocks Green to under 200 engines in Solihull.

Authors' organisations and addresses

R. A. Dover
Land Rover Ltd
Lode Lane
Solihull
West Midlands B92 8NW
England

F. M. Greguras
Fenwick, Davis & West
Two Palo Alto Square
Palo Alto, CA 94306
USA

T. E. Hanley
Land Rover Ltd
Lode Lane
Solihull
West Midlands B92 8NW
England

L. Bisiach
Bisiach & Carru SpA
Corso Lombardi 21
10078 Venaria (Torino)
Italy

G. Repetto
Fata New Hunter
Engineering SpA
Divisione Fata Industriale
Via Traversella 11
10148 Torino
Italy

T. J. Vickerstaff
Jaguar Cars Ltd
Browns Lane
Allesley
Coventry CV5 9DR
England

R. Svensson
Asea Robotics AB
Dept JSV
S 721-83 Vasteras
Sweden

T. Kudo
Mechatronics Development
Department
Nissan Motor Co. Ltd
2-34 Namamugi 2-Chome
Tsurumi-ku
Yokohama-shi
Kanagawa-ken
Japan

W. J. P. A. Verbeek
Industrial Imaging
nv Optische Industrie De Oude
Delft (Oldelft)
PO Box 72
2600 MD Delft
The Netherlands

P. G. Davey
Meta Machines Ltd
9 Blacklands Way
Abingdon Industrial Park
Abingdon
Oxford OX14 1DY
England

E. D. Lawley
Gaydon Technology Ltd
Banbury Road
Lighthorne
Warwick CV35 0BL
England

P. Johnstone
Automatix International (UK) Ltd
Barclays Venture Centre
University of Warwick Science Park
Coventry CV4 7EZ
England

A. D. Wiley
Gaydon Technology Ltd
Banbury Road
Lighthorne
Warwick CV35 0BL
England

H. F. Pitcher
Haden Drysys International
Swan Centre
1506–1508 Coventry Road
Yardley
Birmingham B25 8AD
England

R. Mills
Gaydon Technology Ltd
Banbury Road
Lighthorne
Warwick CV35 0BL
England

D. German
T & C Division
Ford Motor Company
29500 Plymouth Road
Livonia, MI 48150
USA

D. W. Whitney
CPC-Pontiac, Engine Operations,
Manufacturing Engineering
1 Pontiac Plaza
Pontiac, MI 48058-3486
USA

V. Nicolo
Fiat Auto SpA
Direzione Produzione
Pianificazione e G.I.P.
Torino
Italy

M. Guichard
Centre Technique Citröen
Chemin Vicinal 2
F-718140 Velizy
France

G. Nordsten
Engine Division
Volvo Komponenter AB
54187 Skovde
Sweden

J. Lindgren
Saab Automation AB
Box 1017
S-551 11 Jonkoping
Sweden

P. Vicentini
Alfa Romeo Auto SpA
Direzione di Milano
CAP 20149
Italy

L. Dittrich
Schindler-Digitron AG
Erlenstrasse 32
CH-2555 Brugg-Biel
Switzerland

Y. Matsuyama
Ishikawajima-Harima Heavy Industries Co. Ltd
Ihi Kohto Office
19-10, Mohri 1-chome
Kohto-Ku
Tokyo
Japan

I. P. Langley
Decision Support Systems
Istel Ltd
Highfield House
Headless Cross Drive
Headless Cross
Redditch
Worcs B97 5EU
England

Source of material

The management of technological change in the vehicle industry
First presented at the 9th British Robot Association Annual Conference, 13–14 May 1986, Stratford-upon-Avon, UK.
Reprinted Courtesy of the author and the British Robot Association.

Legal considerations in procuring factory automation equipment
Not previously published.

Land Rover robot welding line
First presented at the 2nd International Conference on Robots in the Automotive Industry, 15–17 May 1985, Birmingham, UK.
Reprinted courtesy of the author and IFS (Conferences) Ltd.

Tauro system
First presented at the 2nd International Conference on Robots in the Automotive Industry, 15–17 May 1985, Birmingham, UK.
Reprinted courtesy of the author and IFS (Conferences) Ltd.

Flexible automated body assembly
First presented at the 2nd International Conference on Robots in the Automotive Industry, 15–17 May 1985, Birmingham, UK.
Reprinted courtesy of the author and IFS (Conferences) Ltd.

Bonnet assembly cell at Jaguar Cars
First presented at ISATA '85, 23–27 September 1985, Graz, Austria.
Reprinted courtesy of the author and ISATA.

Car body assembly with ASEA 3D-vision
First presented at the 15th International Symposium on Industrial Robots, 11–13 September 1985, Tokyo, Japan.
Reprinted courtesy of the author and the Japan Industrial Robot Association.

Automated assembly systems in the automotive industry
First presented at the 7th International Conference on Assembly Automation, 4–6 February 1986, Zurich, Switzerland.
Reprinted courtesy of the author and IFS (Conferences) Ltd.

Automatic brazing of car bodies by intelligent robots
First presented at the 2nd International Conference on Machine

Intelligence, 26–28 November 1985, London, UK.
Reprinted courtesy of the authors and IFS (Conferences) Ltd.

Control strategy for an arc-welding robot under sensory guidance
First presented at the 5th International Conference on Robot Vision and Sensory Controls, 29–31 October 1985, Amsterdam, The Netherlands.
Reprinted courtesy of the authors and IFS (Conferences) Ltd.

Laser sensors for arc-welding robots
Not previously published.

Second generation robots for adhesive bonding and sealing
First presented at the 8th British Robot Association Annual Conference, 14–17 May 1985, Birmingham, UK.
Reprinted courtesy of the author and the British Robot Association.

Stereo vision for robotic seam sealant spraying
First presented at the 2nd International Conference on Robots in the Automotive industry, 15–17 May 1985, Birmingham, UK.
Reprinted courtesy of the author and IFS (Conferences) Ltd.

Advanced paint shop automation
First presented at ISATA '85, 23–27 September 1985, Graz, Austria.
Reprinted courtesy of the author and ISATA.

Robots in the unmanned spraybooth
First presented at the 2nd International Conference on Robots in the Automotive Industry, 15–17 May 1985, Birmingham, UK.
Reprinted courtesy of the author and IFS (Conferences) Ltd.

Visual inspection and calibration test by robot
First presented at the 9th British Robot Association Annual Conference, 13–14 May 1986, Stratford-upon-Avon, UK.
Reprinted courtesy of the author and the British Robot Association.

Vision aided inspection techniques for the automotive industry
First presented at VISION '85, 25–28 March 1985, Detroit, USA. (SME Technical Paper No. M585-243.)
Reprinted courtesy of the author and the Society of Manufacturing Engineers.

Optical gauging of engine crankshafts
First presented at VISION '85, 25–28 March 1985, Detroit, USA. (SME Technical Paper No. M585-249.)
Reprinted courtesy of the author and the Society of Manufacturing Engineers.

Monitoring and diagnostics in automotive manufacturing systems
First presented at the International Congress for Metalworking and Automation, 18–19 September 1985, Hanover, West Germany.
Reprinted courtesy of the author and Verein Deutscher Werkzeug maschinenfabriken eV (VDW).

Industrial use of ultrasonic ranging sensors in robotics
First presented at the 6th International Conference on Robot Vision and Sensory Controls, 3–5 June 1986, Paris, France.
Reprinted courtesy of the authors and IFS (Conferences) Ltd.

The evolution of FMS at Volvo
First presented at the 4th International Conference on Flexible Manufacturing Systems, 15–17 October 1985, Stockholm, Sweden.
Reprinted courtesy of the author and IFS (Conferences) Ltd.

Techniques for automatic identification in FMS installations
First presented at the 4th International Conference on Flexible Manufacturing Systems, 15–17 October 1985, Stockholm, Sweden.
Reprinted courtesy of the author and IFS (Conferences) Ltd.

An FMS for mechanical components
First presented at the 4th International Conference on Flexible Manufacturing Systems, 15–17 October 1985, Stockholm, Sweden.
Reprinted courtesy of the authors and IFS (Conferences) Ltd.

The automated storage and transport of stampings
First presented at the 6th International Conference on Automation in Warehousing, 15–17 October 1985, Stockholm, Sweden.
Reprinted courtesy of the author and IFS (Conferences) Ltd.

Automated warehousing for car storage
First presented at the 6th International Conference on Automation in Warehousing, 15–17 October 1985, Stockholm, Sweden.
Reprinted courtesy of the author and IFS (Conferences) Ltd.

Simulation in advanced manufacturing technology
Not previously published.

A guided vehicle system for the assembly of engines on the move
Not previously published.